PROJECT MANAGEMENT AND TEAMWORK

Karl A. Smith
University of Minnesota

Boston   Burr Ridge, IL   Dubuque, IA   Madison, WI   New York   San Francisco   St. Louis
Bangkok   Bogotá   Caracas   Lisbon   London   Madrid
Mexico City   Milan   New Delhi   Seoul   Singapore   Sydney   Taipei   Toronto

McGraw-Hill Higher Education
*A Division of The **McGraw-Hill** Companies*

PROJECT MANAGEMENT AND TEAMWORK
Copyright © 2000 by The McGraw-Hill Companies, Inc. All rights reserved. Printed in the United States of America. Except as permitted under the United States Copyright Act of 1976, no part of this publication may be reproduced or distributed in any form or by any means, or stored in a database or retrieval system, without the prior written permission of the publisher.

This book is printed on acid-free paper.

3 4 5 6 7 8 9 0 FGR/FGR 0 9 8 7 6 5 4 3 2 1

ISBN 0-07-012296-2

Publisher: *Thomas Casson*
Senior sponsoring editor: *Eric M. Munson*
Marketing manager: *John T. Wannemacher*
Project manager: *Karen J. Nelson*
Production supervisor: *Kari Geltemeyer*
Coordinator freelance design: *Craig E. Jordan*
Compositor: *ElectraGraphics, Inc.*
Typeface: *10/12 Times Roman*
Printer: *Quebecor Printing Book Group/Fairfield*

Library of Congress Cataloging-in-Publication Data

Smith, Karl A.
   Project management and teamwork / Karl A. Smith.
      p. cm. — (McGraw-Hill's BEST—basic engineering series and tools)
   ISBN 0-07-012296-2
     1. Engineering management. 2. Teams in the workplace. 3. Industrial project management. I. Title. II. Series.

TA190 .S63 2000
658.4'04—dc21   99-059497

http://www.mhhe.com

PREFACE

When McGraw-Hill invited me to write a module on project management and teamwork for their BEST series, I thought, What a terrific idea! I had been teaching project management and teamwork courses for seniors in engineering; graduate students in professional master's programs, especially at the University of Minnesota's Center for the Development of Technological Leadership; and participants in short courses in the University of Minnesota's Executive Development Program, government agencies, and private companies. It would not have occurred to me to write a book for first-year students. I immediately embraced the idea and started work.

I've been teaching a course for first-year students at the University of Minnesota for more than 20 years. It evolved into a course titled How to Model It: Building Models to Solve Engineering Problems, which I have been teaching with colleagues and undergraduate student teaching assistants for the past 10 years. We also wrote a book to accompany the course—*How to Model It: Problem Solving for the Computer Age* (Starfield, Smith, and Bleloch, 1994). Since this course makes extensive use of project teams, I know that a book on project management and teamwork is needed.

Teamwork and projects are at the heart of the approach I use in teaching students at all levels, including participants in faculty development workshops. I've learned that it isn't easy for students to work effectively in project teams or for faculty to organize and manage them, but the potential for extraordinary work from teams makes it worth the effort. Also, projects and teamwork are a central part of engineering work in the world outside the classroom.

The first part of this book summarizes the context of engineering and stresses the importance of teamwork. The middle part focuses on the nature of projects and the project manager's role. The last part emphasizes the particulars on scheduling, monitoring, and documentation. Overall, my goals for readers of *Project Management and Teamwork* are the following:

- To understand the dynamics of team development and interpersonal problem solving.
- To identify strategies for accelerating the development of true team effectiveness.
- To understand the critical dimensions of project scope, time, and cost management.
- To understand critical technical competencies in project management.
- To explore a variety of "best practices" including anticipating, preventing, and overcoming barriers to project success.

As you engage with this book, be sure to continually reflect on what you're learning and how you can apply it to the projects and teams you work on each day, in classes, on the job, and in social, professional, and community organizations. An important key to success in projects and teams is to routinely work at a "meta level." That means you are simultaneously thinking about the task and how well the team is working. Talk with others about how the projects and teams you're involved with are going, share successes and insights, and work together to identify and solve team problems. The personal story in the accompanying box describes some of the questions I've grappled with and how I got interested in this project. I encourage you to develop your own stories as you work your way through this book.

One of the messages of the story in the box is the importance of checking a variety of resources to help formulate and solve the problems you encounter. Another message is that, although engineers spend some of their time working alone, engineering is not individual, isolated work. Collaborative problem solving and teamwork are central to engineering. Engineers must learn to solve problems by themselves, of course, but they must also learn to work collaboratively to effectively solve the other 95 percent of the problems they will face as professional engineers. There may be a tendency to think that this 95 percent—this asking questions and search-

---

PERSONAL STORY

I have been involved in engineering, as a student and as a professional, for over 30 years. Frequently I have grappled with the questions, What is the engineering method? Is it applied science? Is it design? As a professor I have struggled with the question, What should my students learn and how should they learn it? These concerns prompted me to address the question, What is the nature of engineering expertise and how can it be developed effectively?

A study conducted by one of my colleagues (Johnson, 1982) provides valuable insight into the activities of engineers. My colleague was hired to collect protocol from engineering experts while they solved difficult problems. Working with a team of professors, he developed a set of difficult and interesting problems, which he took to chief engineers in large companies. In case after case the following scenario was repeated.

The engineer would read the problem and say, "This is an interesting problem."

My colleague would ask, "How would you solve it?"

The engineer would say, "I'd check the engineers on the floor to see if any of them had solved it."

In response, my colleague would say, "Suppose that didn't work."

"I'd assign the problem to one of my engineers to check the literature to see if a solution was available in the literature."

"Suppose that didn't work," retorted my colleague.

"Well, then I'd call my friends in other companies to see if any of them had solved it."

Again my colleague would say, "Suppose that didn't work."

"Then I'd call some vendors to see if any of them had a solution."

My colleague, growing impatient at not hearing a problem solution, would say, "Suppose that didn't work."

At some stage in this interchange, the engineer would say, "Well, gee, I guess I'd have to solve it myself."

To which my colleague would reply, "What percentage of the problems you encounter fall into this category?"

Engineer after engineer replied, "About five percent"!

ing other sources for the solution—is either trivial or else unrelated to engineering. However, working with others to formulate and solve problems and accomplish joint tasks is critical to success in engineering.

ACKNOWLEDGMENTS

Many people deserve credit for guidance in this project. Michael B. Mahler, a graduate student in civil engineering at the University of Minnesota, with whom I've taught and worked on project management for many years, provided enormous insight into the process of what will work for students and was a source of constant support and encouragement. Robert C. Johns co-taught the project management course with me at Minnesota and provided lots of good ideas. Anthony M. Starfield, co-creator of the first-year course, How to Model It, and co-author of the book by the same title encouraged me to use the questioning format of the *How to Model It* book to engage the reader. The five manuscript reviewers provided terrific assistance. Holly Stark and Eric Munson of McGraw-Hill, and Byron Gottfried, Consulting Editor, initiated the idea and provided guidance throughout. A special note of thanks to my daughters, Riawa and Sharla Smith, who helped with the editing and the graphics.

A special acknowledgment goes to Michigan State University, which provided me with a wonderful place to work on this project during my sabbatical leave. Another goes to David and Roger Johnson (whose cooperative learning model provides the theoretical basis for this book) for their great ideas, generosity, and steadfast support.

Most of all I thank the hundreds of students who learned from and with me in project management courses for their patience, perseverance, wonderful suggestions and ideas, and interest and enthusiasm in project management and teamwork.

COMMENTS AND SUGGESTIONS

Please send your comments and suggestions to me at ksmith@tc.umn.edu.

REFERENCES

Johnson, P. E. 1982. Personal communication.

Starfield, Anthony M.; Smith Karl A., and Bleloch, Andrew L. 1994. *How to model it: Problem solving for the computer age.* Edina, MN: Burgess.

Contents

Preface   iii

Chapter 1
Project Management and Teamwork in Engineering   1

What Is Engineering?   1
Engineering Design   3
Modeling and Engineering   3
Teamwork and Engineering   5
Systems Approach   6
Reflection: Project Management and
   Teamwork in Engineering   8
Questions   9
Exercises   9
References   10

Chapter 2
Teamwork   13

Definition of a Team   14
Types of Learning Teams   15
   Pseudo Learning Group   15
   Traditional Classroom Learning Group   16
   Cooperative Learning Group   16
   High-Performance Cooperative Learning
      Group   16
Groups and Teams   17
Importance of Diversity   17
Characteristics of Effective Teams   18
Questions   19
Exercises   20
References   20

Chapter 3
Teamwork Skills and Problem Solving   23

Importance of Task and Relationship   23
Organization: Group Norms   24
Communication   26

Leadership   27
Decision Making   29
Conflict Management   32
Teamwork Challenges and Problems   35
Reflection: Teamwork   36
Questions   37
Exercises   37
References   41

Chapter 4
Project Management Principles and Practices   43

What Is a Project?   44
Keys to Project Success   45
Project Life Cycle   47
Project Planning   49
Reflection: Project Management   50
Questions   50
Exercise   51
   Project Planning   51
References   51

Chapter 5
Project Manager's Role   53

Changes in the Workplace   53
Changes in Project Management   56
Skills Necessary for Effective Project
   Managers   58
Project Manager's Role over the Project Life
   Cycle   59
   Planning   59
   Organizing   60
   Staffing   60
   Directing   61
   Controlling   61
Questions   62
Exercise   62
References   63

Chapter 6
PROJECT SCHEDULING 65

Work Breakdown Structure   66
Critical Path Method   67
    Forward Pass—Early Start (ES) and Early
       Finish (EF) 68
    Backward Pass—Late Start (LS) and Late
       Finish (LF) 68
    Critical Path   69
    Floats   69
    Gantt Charts and CritPath   70
    Critical Path Method Summary   72
    Bus Shelter Construction Example   72
Project Resource and Cost Considerations   74
    Resource Leveling   74
    Cost Considerations   76
The Role of Computer-based Project
   Management Software   78
Questions   78
Exercises   78
Reference   80

Chapter 7
PROJECT MONITORING AND EVALUATION 81

Meetings   81
Monitoring Group Effectiveness   83
Team Talk Analysis   86
Project Evaluation   89
    Continual Evaluation   90
Building Quality into Projects   90
Questions   92
References   92

Chapter 8
PROJECT MANAGEMENT DOCUMENTATION AND COMMUNICATIONS 95

Project Documentation   95
Journals and Notes   97
Project Communications   99
    Getting Started, Keeping Going   100
    Revising and Refining   100
Questions   100
References   101

Chapter 9
PROJECT MANAGEMENT SOFTWARE 103

Personal Information Managers   103
Project Management Software   104
Project Management and the World Wide
   Web   108
Questions   109
References   109

Chapter 10
WHERE TO GO FROM HERE 111

References   113

INDEX 115

chapter

1

PROJECT MANAGEMENT AND TEAMWORK IN ENGINEERING

Project management and teamwork are rapidly gaining importance in engineering. Because teamwork and projects in engineering are receiving more and more emphasis in business, industry, and government, they are also becoming common in engineering classes. In addition, teams are used in classes because students working in well-structured teams learn more, remember it longer, and develop problem-solving skills more fully than students working individually.

WHAT IS ENGINEERING?

Periodically throughout this book, I'll ask you to stop and reflect. I encourage you to take advantage of the opportunity. My goals are to give you a chance to describe what you know and to get you to think, so that when you read what I have to say about a topic you'll have a basis for comparing and contrasting new information with information you already have.

Before you read ahead for various answers to the question "What is engineering?" Complete the following Reflection.

REFLECTION   What is engineering, and what does it mean to learn to engineer in school? What is your experience with engineering? Did you learn about engineering in high school? Do you have a brother or sister, mother or father, or other relative or friend who is an engineer? Take a minute to reflect on where you learned about engineering and what your impressions of engineering are.

What did you come up with?

Since there are few high school courses in engineering, it is relatively difficult for first-year college students to have gained much exposure to engineering. Yet we are surrounded by engineering accomplishments; they are so ubiquitous that we don't notice most of them. One of the foremost thinkers and writers on engineering, mechanical engineering professor Billy Koen, is noted for asking four probing questions of his audiences (Koen, 1984). The first is:

1. Can you name one thing in the room in which you are sitting (excluding yourself, of course) that was not developed, produced, or delivered by an engineer?

Koen finds that the question is usually greeted with bewildered silence. I have posed his questions to hundreds of first-year students, and they have come up with some great suggestions: the air (but how does it get into the room?), dirt (trapped in people's shoes), electromagnetic radiation (but the lights generate much more than the background). Almost everything that we encounter was developed, produced, or delivered by an engineer or engineers.

Koen's second question is:

2. Can you name a profession that is affecting your life more incisively than engineering?

Again, students name several professions but on reflection note that if it were not for engineering, politicians would have a difficult time spreading their ideas; doctors, without their tools, would be severely limited in what they could do; lawyers wouldn't have much to read; and so forth. Things such as telephones, computers, airplanes, and skyscrapers—which have an enormous effect on our lives—are all products of engineering.

Koen's third question is:

3. Since engineering is evidently very important, can you now define the engineering method for solving a problem?

Many students respond with a puzzled look, as if I am asking an unfair question. They note that they have a ready response to the question "What is the scientific method?" Students list things like "applied science," "problem solving," and "trial and error," but almost no one (over the 15 or so years that I've been asking this question) says "design."

If you were to ask practicing engineers the question "What is the engineering method?" they would likely respond "Engineering is design!" A group of national engineering leaders has said:

> Design in a major sense is the essence of engineering; it begins with the identification of a need and ends with a product or system in the hands of a user. It is primarily concerned with synthesis rather than the analysis which is central to engineering science. Design, above all else, distinguishes engineering from science. (Hancock, 1986)

We'll explore the concept of engineering design next—and save Koen's fourth and final question for the end of the chapter.

ENGINEERING DESIGN

If design is the essence of engineering, the next question is, What is design? The Accreditation Board for Engineering and Technology (ABET 1999), the group that accredits engineering programs, has defined engineering design as "the process of devising a system, component or process to meet a desired need."

Researchers who carefully observe the engineering design process are increasingly noting that it is quite different from the formal process typically described in textbooks. For example, Eugene Ferguson (1992) writes:

> Those who observe the process of engineering design find that it is not a totally formal affair, and that drawings and specifications come into existence as a result of a social process. The various members of a design group can be expected to have divergent views of the most desirable ways to accomplish the design they are working on. As Louis Bucciarelli (1994), an engineering professor who has observed engineering designers at work, points out, informal negotiations, discussions, laughter, gossip, and banter among members of a design group often have a leavening effect on its outcome.

Recent work on engineering design indicates that it is a more social process than we once thought. Larry Leifer (1997) of the Stanford Center for Design Research claims that engineering design is "a social process that identifies a need, defines a problem, and specifies a plan that enables others to manufacture the solutions." Leifer's research shows that design is fundamentally a social activity. He describes practices such as "negotiating understanding," "conserving ambiguity," "tailoring engineering communications for recipients," and "manipulating mundane representations."

If design is the heart of engineering and design is a social process, then it follows that project management and teamwork are essential to engineering. Many problems with engineering result from poor team dynamics and inadequate project management. In fact, Leifer notes, "Design team failure is usually due to failed team dynamics."

A lot has been written about engineering and engineering design. Adams, (1991), Hapgood (1992), and Ferguson (1992), for example, have written for general audiences and their works can give first-year students considerable insight into engineering. Many writers take a modeling approach to helping students learn about the engineering method and how to do engineering design. Two books that emphasize this connection between modeling and design, and extend it substantially, are Papalambros and Wilde (1988) and Chapman, Bahill, and Wymore (1992).

MODELING AND ENGINEERING

Modeling in its broadest sense is the cost-effective use of one thing in place of another for some cognitive purpose (Rothenberg, 1989). A model represents reality for the given purpose; the model is an abstraction of reality, however, in the sense that it cannot represent all aspects of reality. Models are characterized by three essential attributes:

1. *Reference:* A model is *of* something (its *referent*).
2. *Purpose:* A model has an intended cognitive *purpose* with respect to its referent.
3. *Cost-effectiveness:* A model is more *cost-effective* to use for this purpose than the referent itself would be.

A problem that I often give to help students learn about these attributes of modeling involves determining the maximum number of Ping-Pong balls that could fit in the room they're sitting in. First I give them about 20 seconds and ask each person to guess. Next I ask them to work in groups for about 5 to 10 minutes to develop not only a numerical estimate but also a description of the method they use. At this stage students typically model the room as a rectangular box and the ball as a cube. They then determine the number by dividing the volume of the room by the volume of a ball. I ask them what they would do if I gave them the rest of the class period to work on the problem. Sooner or later a student says, "Who cares how many Ping-Pong balls could fit in the room!" I thank that student and report that we can now stop. In any problem that involves modeling, the purpose must be specified. Without knowing the purpose, we don't know how to use the model; the 20-second answer might be good enough.

An essential aspect of modeling is the use of heuristics (Starfield, Smith, and Bleloch, 1994), which may be generally defined as methods or procedures that aid in discovery or problem solving. Although difficult to define, heuristics are relatively easy to identify using the characteristics listed by Koen (1984, 1985):

1. Heuristics do not guarantee a solution.
2. Two heuristics may contradict or give different answers to the same question and still be useful.
3. Heuristics permit the solving of unsolvable problems or reduce the search time to a satisfactory solution.
4. The heuristic depends on the immediate context instead of absolute truth as a standard of validity.

Thus, a heuristic is anything that provides a plausible aid or direction in the solution of a problem but is in the final analysis unjustified, incapable of justification, and fallible. It is used to guide, to discover, and to reveal. Heuristics are also a key part of Koen's definition of the engineering method:

> The engineering method is the use of heuristics to cause the best change in a poorly understood situation within the available resources. (p. 70)

Typical engineering heuristics include (1) rules of thumb and orders of magnitude, (2) factors of safety, (3) circumstances that determine the engineer's attitude toward his or her work, (4) procedures that engineers use to keep risk within acceptable bounds, and (5) rules of thumb that are important in resource allocation.

Models and heuristics will constitute a major part of this book. The critical path method (CPM) is a procedure for modeling complex projects with interdependent activities. Visual representations include Gantt charts and network diagrams. My goal is for you to develop the skills and confidence necessary to organize, manage, participate in, and lead project teams. This goal is consistent with current thinking about the purpose of engineering schools. W. Edwards Deming associate and engineering educator Myron Tribus (1996) summarized the purpose of engineering schools as follows:

> The purpose of a School of Engineering is to teach students to create value through the design of high quality products and systems of production, and services, and to organize and lead people in the continuous improvement of these designs. (p. 25)

Notice that Tribus considers management an integral part of engineering. He also elaborates on the importance of group work for learning to engineer:

> The main tool for teaching wisdom and character is the group project. Experiences with group activities, in which the members of the groups are required to exhibit honesty, integrity, perseverance, creativity and cooperation, provide the basis for critical review by both students and teachers. Teachers will need to learn to function more as coaches and resources and less as givers of knowledge. (p. 25)

TEAMWORK AND ENGINEERING

The importance of teamwork in business and industry is embedded in the concepts of concurrent (or simultaneous) engineering and total quality management. The following quote elaborates on this point:

> In concurrent engineering (CE), the key ingredient is teamwork. People from many departments collaborate over the life of a product—from idea to obsolescence—to ensure that it reflects customers' needs and desires . . . Since the very start of CE, product development must involve all parts of an organization, effective teamwork depends upon sharing ideas and goals beyond immediate assignments and departmental loyalties. Such behavior is not typically taught in the engineering schools of U.S. colleges and universities. For CE to succeed, teamwork and sharing must be valued just as highly as the traditional attributes of technical competence and creativity, and they must be rewarded by making them an integral part of the engineer's performance evaluation. (Shina, 1991 p. 23)

Project management and teamwork are central to engineering. Learning how to organize and manage projects, and to participate effectively in project teams, will not only serve you well in engineering school, where there are lots of group projects, but will also be critical to your success as a professional engineer. The Boeing Company uses the following checklist when considering new employees.

Employer's checklist—Boeing Company

- ✓ A good grasp of these engineering fundamentals:
    - Mathematics (including statistics)
    - Physical and life sciences
    - Information technology
- ✓ A good understanding of the design and manufacturing process (i.e., an understanding of engineering)
- ✓ A basic understanding of the context in which engineering is practiced, including:
    - Economics and business practice
    - History
    - The environment
    - Customer and societal needs
- ✓ A multidisciplinary systems perspective
- ✓ Good communication skills
    - Written
    - Verbal
    - Graphic
    - Listening
- ✓ High ethical standards
- ✓ An ability to think critically and creatively as well as independently and cooperatively
- ✓ Flexibility—an ability and the self-confidence to adapt to rapid/major change
- ✓ Curiosity and a lifelong desire to learn
- ✓ A profound understanding of the importance of teamwork

SOURCE: *ASEE Prism*, December 1996, p. 11.

SYSTEMS APPROACH

In addition to teamwork, another idea emphasized not only in employer checklists like Boeing's but also in the new accreditation criteria for the Accreditation Board for Engineering and Technology is that of systems and the systems approach.

A *system* is a whole that cannot be divided up into independent parts (Ackoff, 1994). Systems are made up of sets of components that work together for a specified overall objective. The systems approach is simply a way of thinking about total systems and their components.

Five basic aspects must be kept in mind when thinking about the meaning of a system: (1) the whole system's objectives and, more specifically, the performance measures of the whole system; (2) the system's environment, including the fixed constraints; (3) the resources of the system; (4) the components of the system, their activities, goals, and measures of performance; and (5) the management of the system (Churchman, 1968).

Systems thinking is a discipline for seeing wholes. It is a framework for seeing interrelationships rather than things, for seeing patterns of change rather than static snapshots. It is a set of principles and a set of specific tools and techniques (Senge, 1990). An implication of the systems approach is that everybody involved must work together to improve whole systems (Weisbord, 1987). The systems approach is commonly operationalized through *learning organizations.* Peter Senge (in Ray and Rinzler, 1993) lists five factors, or disciplines, that make up the art and practice of the learning organization:

1. *Building shared vision.* The idea of *building* shared vision stresses that you never quite finish it—it's an ongoing process.
2. *Personal mastery.* Learning organizations must be fully committed to the development of each individual's personal mastery—each individual's capacity to create a life the way he or she truly wants.
3. *Mental models.* Our vision of current reality has everything to do with the third discipline—*mental models*—because what we really have in our lives is constructions, internal pictures that we continually use to interpret and make sense out of the world.
4. *Team learning.* Individual learning, no matter how wonderful it is or how great it makes us feel, is fundamentally irrelevant to organizations, because virtually all important decisions occur in groups. The learning units of organizations are "teams," groups of people who need one another to act.
5. *Systems thinking.* The last discipline, the one that ties them all together, is *systems thinking.*

As in many other project management books, a systems theme will be one of the integrating themes in this book. The idea of systems, along with that of the learning organization, has important contributions to make not only to your study of project management but also to many other things you will be studying in engineering. Here, for example, are eight principles for learning from the Xerox Corporation (Jordan, 1997 p. 116):

1. Learning is fundamentally social.
2. Cracking the whip stifles learning.
3. Learning needs an environment that supports it.
4. Learning crosses hierarchical bounds.
5. Self-directed learning fuels the fire.
6. Learning by doing is more powerful than memorizing.
7. Failure to learn is often the fault of the system, not the people.
8. Sometimes the best learning is unlearning.

The above list indicates that the ideas in this book are important not only for your project work but also for your day-to-day work in engineering school.

A 1998 survey indicated that design is the first and management is the third most frequent work activity reported by engineers (see Table 1.1).

Table 1.1 Rank order of work activities, 1993

Activity	Percent Mentioning
1. Design	66%
2. Computer applications	58
3. Management	49
4. Development	47
5. Accounting, etc.	42
6. Applied research	39
7. Quality or productivity	33
8. Employee relations	23
9. Sales	20
10. Basic research	15
11. Production	14
12. Professional services	10
13. Other work activities	8
14. Teaching	8

SOURCE: Burton, Parker, and LeBold, 1998, p. 19.

REFLECTION: PROJECT MANAGEMENT AND TEAMWORK IN ENGINEERING

As I finished writing this book, I was reminded of a book I read almost 20 years ago—*Excellence in Engineering,* by W. H. Roadstrum (1978). The second edition is titled *Being Successful as an Engineer* (Roadstrum, 1988). In this edition, Roadstrum remarks, "Engineering is almost completely divorced from this concept of routine and continuous. Engineering work is project work" (p. 7). Engineering work *is* project work! This is the essence of Roadstrum's book. The first two chapters, "What Engineering Is" and "The Engineer," cover ground similar to the material given in this chapter, but from a perspective about 12 years ago. Roadstrum then addresses "The Project and the Project Team" and "Project Control" in Chapters 3 and 4. Although I had not looked at Roadstrum's book for several years, I was struck by the overlaps between his book and mine.

Being Successful as an Engineer addresses a broad range of topics, including problem solving, laboratory work, design, research and development, manufacturing and quality control, systems, proposal work, human relations, and creativity. Roadstrum writes, "Design is the heart of the engineering process—its most characteristic activity." Furthermore, he states, "If you and I are going to understand engineering, we'll have to understand design" (p. 97).

Roadstrum elaborates on the role of the project engineer, with the following statement:

> Every engineer looks forward to the time when he can have a project of his own. A project engineer has the best job in the business. He has ultimate responsibility for the work as a whole. He is the real architect of the project solution. Even more than his colleagues, he looks at the job as a whole from the beginning. He watches carefully to make all details come together into a timely, economical, fresh, and effective meeting of the need. (p. 166)

Roadstrum's book and ideas no doubt influenced my decision to develop skills and expertise in project management; however, the specific reference lay dormant until now. I hope this book will influence your experience and practice of project management in engineering.

As a final note, recall the discussion from the beginning of the chapter of Professor Billy Koen's probing questions. Koen's fourth question is, "Lacking a ready answer [to the third question—What is the engineering method?], can you then name a nationally known engineer who is wise, well read, and recognized as a scholar in the field of engineering—one to whom I can turn to find out what engineering really is?" To whom would you turn? Difficult, isn't it? No other profession lacks knowledgeable, clearly recognized spokespersons. I sincerely hope that you'll help provide the leadership to make engineering better known.

QUESTIONS

1. What is engineering? How does engineering differ from science? What role does design play in engineering?
2. What is a model? Why are models useful in project management and in engineering?
3. What is a system? Why are many project management books organized around a systems approach?

EXERCISES

1. Summarize your course work and experiences with engineering and design. What are some of the key things you've learned about engineers and engineering? Do you have relatives or friends who are project managers or engineers? If so, talk with them.
2. Why should you, as a first-year engineering student, be interested in project management and teamwork? Take a minute and reflect. Jot down at least three reasons why a first-year engineering student should be interested in project management and teamwork. What did you come up with? Did you say, for instance, that project management and teamwork are integral to engineering professional practice?
3. List your good experiences with projects and teamwork. Have you had experience with a team that had extraordinary accomplishments? If so, describe the

situation, especially the characteristics of the team that led to its success. What were some of the factors—a sense of urgency? a project too complex or timeline too short for one person to complete? or a need for synergistic interaction?

REFERENCES

Ackoff, Russell L. 1994. *The democratic corporation: A radical prescription for recreating corporate America and rediscovering success.* Oxford: Oxford University Press.

Adams, James L. 1991. *Flying buttresses, entropy, and o-rings: The world of an engineer.* Cambridge, MA: Harvard University Press.

ABET. 1999. www.abet.org

Bucciarelli, Louis. 1994. *Designing engineers.* Cambridge, MA: MIT Press.

Burton, Lawrence, Linda Parker, and William K. LeBold. 1998. U.S. engineering career trends. *ASEE Prism,* 7(9): 18–21.

Chapman, William L., A. Terry Bahill, and A. Wayne Wymore. 1992. *Engineering modeling and design.* Boca Raton, FL: CRC Press.

Churchman, C. West. 1968. *The systems approach.* New York: Laurel.

Ferguson, Eugene S. 1992. *Engineering and the mind's eye.* Cambridge, MA: MIT Press.

Hancock, J. C., Chairman (1986). *Workshop on undergraduate engineering education.* Washington, DC: National Science Foundation.

Hapgood, Fred. 1992. *Up the infinite corridor: MIT and the technical imagination.* Reading, MA: Addison-Wesley.

Jordan, Brigitte. 1996. 8 principles for learning. *Fast Company* 5 (October/November): 116.

Koen, Billy V. 1984. Toward a definition of the engineering method. *Engineering Education* 75: 151–155.

———. 1985. *Definition of the engineering method.* Washington: American Society for Engineering Education.

Leifer, Larry. 1997. A collaborative experience in global product-based learning. November 18, 1997. National Technological University Faculty Forum.

Papalambros, Panos Y., and Douglass J. Wilde. 1988. *Principles of optimal design: Modeling and computation.* Cambridge, England: Cambridge University Press.

Ray, Michael, and Alan Rinzler, eds. 1993. *The new paradigm in business: Emerging strategies for leadership and organizational change.* Los Angeles: Tarcher/Perigee.

Roadstrum, W. H. 1988. *Being successful as an engineer.* San Jose: Engineering Press.

Roadstrum, W. H. 1978. *Excellence in engineering.* New York: Wiley.

Rothenberg, Jeff. 1989. The nature of modeling. In *Artificial intelligence, simulation & modeling,* edited by L. E. Widman, K. A. Loparo, and N. R. Nielsen. New York: Wiley: pp. 75–90.

Senge, Peter. 1990. *The fifth discipline: The art and practice of the learning organization.* New York: Doubleday.

Shina, S. G. 1991. New rules for world-class companies. Special Report on Concurrent Engineering, edited by A. Rosenblatt and G. F. Watson. *IEEE Spectrum* 28(7): 22–37.

Tribus, Myron. 1996. Total quality management in schools of business and engineering. In *Academic initiatives in total quality for higher education,* edited by Harry V. Roberts, 17–40. Milwaukee: ASQC Quality Press.

Weisbord, Marvin R. 1987. *Productive workplaces: Organizing and managing for dignity, meaning, and community.* San Francisco: Jossey-Bass.

chapter 2

TEAMWORK

Everyone has to work together; if we can't get everybody working toward common goals, nothing is going to happen.

Harold K. Sperlich
Former President, Chrysler Corporation

Coming together is a beginning;
Keeping together is progress;
Working together is success.

Henry Ford

REFLECTION Think about a really effective team that you've been a member of, a team that accomplished extraordinary things and perhaps was even a great place to be. Start by thinking about teams in an academic, professional, or work setting. If no examples come to mind, then think about social or community-based teams. If again you don't conjure up an example, then think about sports teams. Finally, if you don't come up with a scenario from any of these contexts, then simply imagine yourself as a member of a really effective team. OK, got a picture of the team in mind? As you recall (or imagine) this highly effective team experience, try to extract the specific characteristics of the team. What was it about the team that made it so effective? Please make a list.

Look over the list you made in the above Reflection. Did you preface your list with "It depends"? The characteristics of an effective team depend, of course, on the purpose of the team. In large measure, they depend on the team's task goals (those concerning what the team is to do) and maintenance goals (those concerning how the team functions). Michael Schrage (1991) states emphatically:

[P]eople should understand that real value in the sciences, the arts, commerce, and, indeed one's personal and professional lives, comes largely from the process of collaboration. What's more, the quality and quantity of meaningful collaboration often depend upon the tools used to create it . . . Collaboration is a *purposive* relationship. At the heart of collaboration is a desire or need to: solve a problem, create, or discover something. (p. 34) Within a set of constraints—expertise, time, money, competition, conventional wisdom. (p. 36)

Let's assume that it's a team that has both task and maintenance goals, since most effective teams not only have a job to do (a report to write, a project to complete, a presentation to give, etc.) but also a goal of getting better at working with one another.

I've used the Reflection above with hundreds of faculty and students in workshop and classroom settings. Here is a typical list of the characteristics of effective teams:

Good participation	Common goal
Respect	Sense of purpose
Careful listening	Good meeting facilitation
Leadership	Empowered members
Constructively managed conflict	Members take responsibility
Fun, liked to be there	Effective decision making

DEFINITION OF A TEAM

Katzenbach and Smith (1993) studied teams that performed at a variety of levels and came up with four categories. *Pseudo teams* are those that perform below the level of the average member. *Potential teams* don't quite get going but struggle along at or slightly above the level of the average member. *Real teams* perform quite well, and *high-performing teams* perform at an extraordinary level. Katzenbach and Smith then looked for common characteristics of real teams and high-performing teams. All real teams could be defined as follows: a small number of people with complementary skills who are committed to a common purpose, performance goals, and an approach for which they hold themselves mutually accountable. High-performing teams met all the conditions of real teams and, in addition, had members who were deeply committed to one another's personal growth and success.

REFLECTION   Now think about the groups that are being used in your engineering classes. Think about your most successful or effective group project experience. What were the characteristics of the group? What were the conditions? Are they similar to your most effective groups?

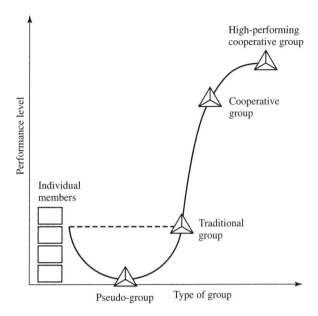

Figure 2.1 Group performance

TYPES OF LEARNING TEAMS

There is nothing magical about teamwork in engineering classes. For example, while some types of learning teams increase the quality of classroom life and facilitate student learning, others hinder student learning and create disharmony and dissatisfaction with classroom life. To use teamwork effectively, you must know what is and what is not a desirable characteristic.

When you choose to use (or are asked or required to use) instructional groups, you must ask yourself, "What type of group am I involved in?" Figure 2.1 and the following may be helpful in answering that question.

PSEUDO LEARNING GROUP

Students in a pseudo learning group are assigned to work together but they have no interest in doing so. They believe they will be evaluated by being ranked from the highest performer to the lowest performer. While on the surface students talk to each other, under the surface they are competing. Because they see each other as rivals who must be defeated, they block or interfere with each other's learning, hide information from each other, attempt to mislead and confuse each other, and distrust each other. Students would achieve more if they were working alone.

Traditional Classroom Learning Group

Students in a traditional classroom learning group are assigned to work together and accept that they must do so. Assignments are structured, however, so that very little joint work is required. Students believe that they will be evaluated and rewarded as individuals, not as members of the group. They interact primarily to clarify how assignments are to be done. They seek each other's information, but have no motivation to teach what they know to their groupmates. Helping and sharing are minimized. Some students loaf, seeking a free ride on the efforts of their more conscientious groupmates. The conscientious members begin to feel exploited and therefore do less. The sum of the whole is more than the potential of some of the members, but the more hard-working and conscientious students would perform higher if they worked alone.

Cooperative Learning Group

Students in cooperative learning groups are assigned to work together and, given the complexity of the task and the necessity for diverse perspectives, they are relieved to do so. They know that their success depends on the efforts of all group members. The group format is clearly defined. First, the group goal of maximizing all members' learning provides a compelling common purpose that motivates members to roll up their sleeves and accomplish something beyond their individual achievements. Second, group members hold themselves and each other accountable for doing high-quality work to achieve their mutual goals. Third, group members work face-to-face to produce joint results. They do real work together. Students promote each other's success through helping, sharing, assisting, explaining, and encouraging. They provide both academic and personal support based on their commitment to and concern for each other. Fourth, group members are taught teamwork skills and are expected to use them to coordinate their efforts and achieve their goals. Both task and maintenance (team-building) skills are emphasized. All members share responsibility for providing leadership. Finally, groups analyze how effectively they are achieving their goals and how well members are working together. There is an emphasis on continual improvement of the quality of learning and teamwork processes. A recent guide to success in active learning is available in the book *Striving for Excellence in College* (Browne and Keeley, 1997).

High-Performance Cooperative Learning Group

A high-performance cooperative learning group meets all the criteria for being a cooperative learning group and outperforms all reasonable expectations, given its membership. What differentiates the high-performance group from the ordinary cooperative learning group is the level of commitment members have to each other and to the group's success. Jennifer Futernick, who is part of a high-performing, rapid-response team at McKinsey & Company, calls the emotional binding together of her teammates a form of love (Katzenbach and Smith, 1993). Ken Hoepner of the

Burlington Northern Intermodal Transport Team (also described in Katzenbach and Smith) stated: "Not only did we trust each other, not only did we respect each other, but we gave a damn about the rest of the people on this team. If we saw somebody vulnerable, we were there to help." Members' mutual concern for each other's personal growth enables high-performance cooperative groups to perform far above expectations, and also to have lots of fun. The bad news about extraordinarily high-performance cooperative learning groups is that they are rare. Most groups never achieve this level of development.

GROUPS AND TEAMS

I've been using the terms *group* and *team* interchangeably and I will continue to do so throughout this book. The traditional literature focuses on groups, while recently some writers have been making distinctions between groups and teams. Katzenbach and Smith (1993) summarize the major differences between working groups and teams (see Table 2.1).

Table 2.1 Not all groups are teams: How to tell the difference

Working Group	Team
Strong, clearly focused leader	Shared leadership roles
Individual accountability	Individual and mutual accountability
Purpose the same as the broader organizational mission	Specific team purpose that the team itself delivers
Individual work-products	Collective work-products
Runs efficient meetings	Encourages open-ended discussion and active problem-solving meetings
Measures its effectiveness indirectly by its influence on others	Measures performance directly by assessing collective work-products
Discusses, decides, and delegates	Discusses, decides, and does real work together

SOURCE: Katzenbach and Smith, 1993.

From your perspective, are there any surprises in Table 2.1? Many students emphasize the importance of a strong leader, but Katzenbach and Smith indicate that real teams, as opposed to working groups, have shared leadership roles. Also notice that the literature on high-performance teams indicates that they are composed of members with complementary skills; that is, they're diverse.

IMPORTANCE OF DIVERSITY

Often we must work with people who are different from us or difficult to work with but whose skills, talents, expertise, and experience are essential to the project. Working with a diverse group may seem impossible at times, but look at the example of Phil Jackson, former head coach of the Chicago Bulls basketball team. Can you imagine more a more diverse group than one made up of Dennis Rodman, Michael Jordan, and Scottie Pippin? Phil Jackson is an expert at managing diversity. Ethnic diversity is increasing in the workplace and in the broader society. Many predict that today's ethnic minorities will grow as a proportion of the population; in fact, Hispanics are predicted to become the majority in the near future.

Diversity has many faces, including preferred learning style (visual, auditory, kinesthetic); social background and experience; ethnic and cultural heritage; gender; and sexual orientation. The evidence from effective groups is that diversity is important; that is, the better a group represents the broader community, the more likely it is to make significant, creative, and acceptable contributions. Participating in and managing diverse groups are not always easy tasks, since diverse groups usually encompass a wide range of ideas and priorities. The following are some steps you can take in learning to manage diverse groups more effectively (Cabanis, 1997; Cherbeneau, 1997):

1. Learn skills for working with all kinds of people.
2. Stress that effective teams are diverse.
3. Stress the importance of requirements.
4. Emphasize performance.
5. Develop perspective-taking skills, i.e. putting yourself in other's shoes.
6. Respect and appreciate alternative perspectives.

The Chicago Bulls' former head coach Phil Jackson has said, "Good teams become great ones when the members trust each other enough to surrender the 'me' for the 'we.'" His 1995 book (coauthored with Hugh Delehanty), *Sacred Hoops: Spiritual Lessons of a Hardwood Warrior,* offers terrific advice on organizing and managing extraordinarily high-performing teams.

CHARACTERISTICS OF EFFECTIVE TEAMS

The research on highly effective teams both in the classroom (Johnson, Johnson, and Smith, 1991, 1998a, 1998b) and in the workplace (Bennis and Biederman, 1997; Hargrove, 1998; Katzenbach and Smith, 1993; Schrage, 1991, 1995) reveals a short list of the characteristics of effective teams:

1. *Positive interdependence.* The group focuses on a common goal or single product.
2. *Individual and group accountability.* Each person takes responsibility for both his or her work and the overall work of the group.
3. *Promotive interaction.* The members do real work, usually face-to-face.

4. *Teamwork skills.* Each member has and practices effective communication (especially careful listening), decision making, problem solving, conflict management, and leadership.
5. *Group processing.* The group periodically reflects on how well the group is working, celebrates the things that are going well, and problem-solves the things that aren't.

Teams have become commonplace in engineering practice and are making inroads in engineering education. The immense literature on teams and teamwork, ranges from very practical guides (e.g., Scholtes, Joiner, and Streibel, 1996; Brassand, 1995) to conceptual and theoretical treatises (e.g., Johnson and Johnson, 1991; Hackman, 1990). Check out one of these to broaden and deepen your understanding of teamwork. Four books were highlighted in this chapter—*Shared Minds: The New Technologies of Collaboration* (Schrage, 1991); *The Wisdom of Teams: Creating the High-Performance Organization* (Katzenbach and Smith, 1993); *Organizing Genius: The Secrets of Creative Collaboration* (Bennis and Biederman, 1997); and *Mastering the Art of Creative Collaboration* (Hargrove, 1998). These four books focus on extraordinary teams—teams that perform at unusually high levels and whose members experience accomplishments through synergistic interaction that they rarely experience in other settings. They provide lots of examples and insights into high-performance teams. Katzenbach and Smith, for example, give the following advice for building team performance:

- Establish urgency and direction.
- Select members based on skill and potential, not personalities.
- Pay attention to first meeting and actions.
- Set clear rules of behavior.
- Set some immediate performance-oriented tasks and goals.
- Challenge the group regularly with fresh information.
- Spend *lots* of time together.
- Exploit the power of positive feedback, recognition, and reward.

Effective teamwork is not easy to accomplish. Engineering professor Douglas J. Wilde said, "It's the soft stuff that's hard, the hard stuff is easy." (Leifer, 1997) However if you work at it, and continue to study and learn about effective teamwork, you will very likely have many positive team experiences (and save yourself a lot of grief). Chapter 3 presents specific skills and strategies needed for effective teamwork.

QUESTIONS

1. What are the characteristics of effective teams? How do you help promote them?

2. Where and how have teamwork skills been taught or emphasized to you? In school? social groups? professional groups? your family? Describe two or three instances where teamwork skills were emphasized.

3. How is increasing ethnic diversity affecting project teams? What are some strategies for effectively participating on and managing diverse teams?

4. Students often remark, "But groups in school are different from groups in the workplace." The remark is delivered as a reason for not using groups in school. Is it a valid excuse? Summarize the major differences between groups in school and groups in the workplace. How are these differences beneficial or harmful to the work of the group? What are some things that you can do to improve the school groups?

Exercises

1. Check out a study of teams that have performed at extraordinary levels. Some of the books listed in the references for this chapter have terrific stories of stellar teams. You may want to check the library or do an electronic search of the literature. Summarize the features of extraordinary teams. How do they compare with the list provided in this chapter? Remember, this is a dynamic area of research with lots of new books and articles appearing each year.

2. Look for opportunities to participate on a superb team. Make a plan for participating on a high-performance team.

3. Study the diversity of teams in your school or workplace, and note strategies for recognizing, valuing, and celebrating diversity.

References

Bennis, Warren, and Patricia Biederman. 1997. *Organizing genius: The secrets of the creative organization.* Reading, MA: Addison-Wesley.

Brassand, Michael. 1995. *The team memory jogger: A pocket guide for team members.* Madison, WI: GOAL/QPC and Joiner Associates.

Browne, M. Neil, and Stuart Keeley. 1997. *Striving for excellence in college.* Upper Saddle River, NJ: Prentice-Hall.

Cabanis, Jeannette. 1997. Diversity: This means you. *PM Network* 11(10): 29–33.

Cherbeneau, Jeanne. 1997. Hearing every voice: How to maximize the value of diversity on project teams. *PM Network* 11(10): 34–36.

Hackman, J. R. 1990. *Groups that work (and those that don't): Creating conditions for effective teamwork.* San Francisco: Jossey-Bass.

Hargrove, Robert. 1998. *Mastering the art of creative collaboration.* New York: McGraw-Hill.

Jackson, Phil, and Hugh Delehanty. 1995. *Sacred hoops: Spiritual lessons of a hardwood warrior.* Hyperion.

Johnson, David W., and Frank P. Johnson. 1991. *Joining together: Group theory and group skills,* 4th ed. Englewood Cliffs, NJ: Prentice-Hall.

Johnson, David W., Roger T. Johnson, and Karl A. Smith. 1991. *Cooperative learning: Increasing college faculty instructional productivity,* Washington: ASHE-ERIC Reports on Higher Education.

———. 1998a. *Active learning: Cooperation in the college classroom,* 2nd ed. Edina, MN: Interaction Book Company.

———. 1998b. Maximizing instruction through cooperative learning. *ASEE Prism* 7(6): 24–29.

Katzenbach, Jon, and Douglas Smith. 1993. *The wisdom of teams: Creating the high-performance organization.* Cambridge, MA: Harvard Business School Press.

Leifer, L. 1997. Design team performance: Metrics and the impact of technology. In S. M. Brown & C. J. Seidner, eds., *Evaluating corporate training: Models and issues.* Kluwer Academic Publishers: 297–320.

Scholtes, Peter R., Brian L. Joiner, and Barbara J. Streibel. 1996. *The team handbook,* 2nd ed. Madison, WI: Joiner Associates.

Schrage, Michael. 1991. *Shared minds: The new technologies of collaboration.* New York: Random House.

———. 1995. *No more teams! Mastering the dynamics of creative collaboration.* New York: Doubleday.

chapter

3

TEAMWORK SKILLS AND PROBLEM SOLVING

I will pay more for the ability to deal with people than any other ability under the sun.

John D. Rockefeller

If you can't operate as a team player, no matter how valuable you've been, you really don't belong at GE.

John F. Welch
CEO, General Electric

REFLECTION   Have you been a member of a team that got the job done (wrote the report, finished the project, completed the laboratory assignment) but that ended up with the members hating one another so intensely they never wanted to see each other again? Most students have, and they find it very frustrating. Similarly, have you been a member of a team whose members really enjoyed one another's company and had a great time socially, but in the end hadn't finished the project? Again, most students have been a member of this type of group and they find it also a frustrating experience. Take a moment to recall your experiences with these two extremes of teamwork.

IMPORTANCE OF TASK AND RELATIONSHIP

As noted in Chapter 2, to be most effective, groups need to do two things very well: accomplish the task and get better at working with one another. Both of these require leadership—not just from a single person acting as the leader but also from every member contributing to the leadership of the group. This chapter focuses on teamwork skills using a "distributed actions approach" to leadership. *Distributed actions* are specific behaviors that group members engage in to help the group accomplish its task or to improve working relationships. Napier and Gershenfeld (1973) summarize many of these behaviors (see Table 3.1). Note the date—1973—which indicates that effective group work is not a new concept.

Table 3.1 Group task and maintenance roles

Group Task Roles	Group Maintenance Roles
Initiating	Encouraging
Seeking information	Expressing feelings
Giving information	Harmonizing
Seeking opinions	Compromising
Giving opinions	Facilitating communications
Clarifying	Setting standards or goals
Elaborating	Testing agreement
Summarizing	Following

SOURCE: Napier and Gershenfeld, 1973.

To realize the benefits of a team culture requires a change in management behavior, as shown in Table 3.2. If the behaviors listed on the right-hand side of Table 3.2 are not common in the groups you participate in, read on.

Table 3.2 Management behavior change needed for team culture

From	To
Directing	Guiding
Competing	Collaborating
Relying on rules	Focusing on the process
Using organizational hierarchy	Using a network
Consistency/sameness	Diversity/flexibility
Secrecy	Openness/sharing
Passive acceptance	Risk taking
Isolated decisions	Involvement of others
People costs	People assets
Results thinking	Process thinking

SOURCE: McNeill, Bellamy, and Foster, 1995.

ORGANIZATION: GROUP NORMS

A common way to promote more constructive and productive teamwork is to have the teams create a set of guidelines for the group, sometimes called group norms.

Take a minute and list some things (attitudes, behaviors, and so on) that you have found or think that would help a group be more effective. Then compare your list with the following two lists, both of which are from McNeill, Bellamy, and Foster (1995). The first was adapted from the Boeing Airplane Group's training manual for team members, and the second is from the Ford Motor Company.

Code of Cooperation

1. *Every* member is responsible for the team's progress and success.
2. Attend all team meetings and be on time.
3. Come prepared.
4. Carry out assignments on schedule.
5. Listen to and show respect for the contributions of other members; be an active listener.
6. *Constructively* criticize ideas, not persons.
7. Resolve conflicts constructively.
8. Pay attention; avoid disruptive behavior.
9. Avoid disruptive side conversations.
10. Only one person speaks at a time.
11. Everyone participates; no one dominates.
12. Be succinct; avoid long anecdotes and examples.
13. No rank in the room.
14. Respect those not present.
15. Ask questions when you do not understand.
16. Attend to your personal comfort needs at any time, but minimize team disruption.
17. Have fun.
18. ?

Ten Commandments: An Affective Code of Cooperation

- Help each other be right, not wrong.
- Look for ways to make new ideas work, not for reasons they won't.
- If in doubt, check it out. Don't make negative assumptions about each other.
- Help each other win, and take pride in each other's victories.
- Speak positively about each other and about your organization at every opportunity.
- Maintain a positive mental attitude no matter what the circumstances.
- Act with initiative and courage, as if it all depends on you.
- Do everything with enthusiasm; it's contagious.
- Whatever you want, give it away.

- Don't lose faith.
- Have fun.

Having an agreed-upon code of cooperation such as the ones listed above will help groups get started toward working effectively. However, if group members haven't developed the requisite communication, trust, loyalty, organization, leadership, decision-making procedures, and conflict management skills, then the group will very likely struggle or at least not perform up to its potential. One way a team can develop such a code is to create a *team charter,* which includes the following:

- Team name, membership, and roles.
- Team mission statement.
- Anticipated results (goals).
- Specific tactical objectives.
- Ground rules/guiding principles for team participation.
- Shared expectations/aspirations.

Team charters are typically created during a team meeting early in the project life cycle. Involvement of all team members in creating the charter helps build commitment of each to the project and other team members. A set of guidelines such as those listed above often help the team through this process.

COMMUNICATION

Effective communication—listening, presenting, persuading—is at the heart of effective teamwork. The task and maintenance roles listed above all involve oral communication. Here are the listening skills emphasized in Arizona State University's Introduction to Engineering Design (McNeill, Bellamy & Foster, 1995):

Stop talking.

Engage in one conversation at a time.

Empathize with the person speaking.

Ask questions.

Don't interrupt.

Show interest.

Concentrate on what is being said.

Don't jump to conclusions.

Control your anger.

React to ideas, not to the speaker.

Listen for what is not said; ask questions.

Share the responsibility for communication.

Three listening techniques they recommend are:

Critical listening
- Separate fact from opinion.

Sympathetic listening.
- Don't talk—listen.
- Don't give advice—listen.
- Don't judge—listen.

Creative listening.
- Exercise an open mind.
- Supplement your ideas with another person's ideas and vice versa.

You may be wondering why so much emphasis on listening. The typical professional spends about half of his or her business hours listening and project managers may spend an even higher proportion of their time listening. Most people, however, are not 100 percent efficient in their listening. Typical listening efficiencies are only 25 percent (Taylor, 1998). The first list provides suggestions to help the listener truly hear what is being said and the second highlights that different situations call for different types of listening.

REFLECTION   Take a moment to think about listening skills and techniques. Do you listen in all three ways listed above? Which are you best at? Which do you need to work on?

LEADERSHIP

A common notion is that leadership is a trait that some people are born with. Another common notion is that a person's leadership ability depends on the situation. There is an enormous literature on leadership, so I'll provide only insights that I've found useful. I'll also try to guide you to more reading and resources on the topic.

INDIVIDUAL AND GROUP REFLECTION   What does it mean to lead a team? What does it take? Take a moment to reflect on the characteristics you admire most in a leader. Jot down 8 to 10 of them. Compare with your team.

Leadership authors Kouzes and Posner (1987, 1993) have asked thousands of people to list the characteristics of leaders they admire. Table 3.3 lists the most common responses from their 1987 and 1993 studies. Many students and workshop participants express surprise at the listing of honesty as the characteristic mentioned most often. They say it's a given. Apparently honesty is not a given for many

Table 3.3 Characteristics of admired leaders

Characteristic	1987 U.S. Percentage of People Selecting	1993 U.S. Percentage of People Selecting
Honest	83%	87%
Forward-looking	62	71
Inspiring	58	68
Competent	67	58
Fair-minded	40	49
Supportive	32	46
Broad-minded	37	41
Intelligent	43	38
Straightforward	34	34
Courageous	27	33
Dependable	32	32
Cooperative	25	30
Imaginative	34	28
Caring	26	27
Mature	23	14
Determined	20	13
Ambitious	21	10
Loyal	21	10
Self-controlled	13	5
Independent	13	5

SOURCE: Kouzes and Posner, 1987, 1993.

leaders in business and industry. In 1993, Kouzes and Posner also asked the respondents to list the most desirable characteristics of colleagues. Honest was number one again, with 82 percent selecting it. Cooperative, dependable, and competent were second, third, and fourth, with slightly over 70 percent of respondents selecting each.

Kouzes and Posner found that when leaders do their best, they challenge, inspire, enable, model, and encourage. They suggest five practices and 10 behavioral commitments of leadership.

Challenging the Process
1. Search for opportunities.
2. Experiment and take risks.

Inspiring a Shared Vision
3. Envision the future.
4. Enlist others.

Enabling Others to Act
  5. Foster collaboration.
  6. Strengthen others.

Modeling the Way
  7. Set the example.
  8. Plan small wins.

Encouraging the Heart
  9. Recognize individual contributions.
  10. Celebrate accomplishments.

Peter Scholtes, author of the best-selling book *The Team Handbook,* also wrote *The Leader's Handbook* (1998). He offers the following six "New Competencies" for leaders:

1. The ability to think in terms of systems and knowing how to lead systems.
2. The ability to understand the variability of work in planning and problem solving.
3. Understanding how we learn, develop, and improve; leading true learning and improvement.
4. Understanding people and why they behave as they do.
5. Understanding the interaction and interdependence between systems, variability, learning, and human behavior; knowing how each affects the others.
6. Giving vision, meaning, direction, and focus to the organization.

In addition to group norms, communication, and leadership, teamwork depends on effective decision making and constructive conflict management, described in the next two sections.

Decision Making

This section on decision making includes both strategies for decision making in groups and more general considerations for addressing ranking tasks.

INDIVIDUAL AND GROUP REFLECTION   How do you typically make decisions in groups? Do you vote? Do you defer to the "expert"? Do you try to reach consensus? Take a moment to reflect on how the groups you participate in typically make decisions.
What did you come up with? Compare your reflection with those of other group members.

There are several approaches to making decisions in groups. Before exploring them, however, I suggest that you try a group decision-making exercise. Common exercises to assist in the development of teamwork skills, especially communication (sharing knowledge and expertise), leadership, and decision making are ranking tasks, such as the survival tasks, in which a group must decide which items are most important for survival in the desert, on the moon, or in some other difficult place. Ranking tasks are common in organizations that must select among alternative designs, hire personnel, or choose projects or proposals for funding.

My favorite ranking task for helping groups focus on communication, leadership, decision making, and conflict resolution is "They'll Never Take Us Alive." This exercise, which includes both individual and group decision making, is included at the end of this chapter. Do it now.

GROUP REFLECTION 1   How did your group make the decision? Did you average your individual rankings? Vote? Did you discuss your individual high and low rankings and then work from both ends toward the middle? Did you try to reach consensus? Were you convinced by group members who seem to have "expert" knowledge? Did you start with the number of fatalities for one of the activities and work from there?

GROUP REFECTION 2   How well did your group work? What went well? What things could you do even better next time?

The method a group uses to make a decision depends on many factors, including how important the decision is, and how much time there is. Groups should have a good repertoire of decision-making strategies and a means of choosing the one that is most appropriate for the situation.

Several methods have been described in the literature for making decisions. One of my favorites is from David Johnson and Frank Johnson (1991). The authors list seven methods for making decisions:

1. *Decision by authority without discussion.* The leader makes all the decisions without consulting the group. It is efficient but does not build team member commitment to the decision.
2. *Expert member.* Group decision made by letting the most expert member decide for the group. The difficulty is often deciding who has the most expertise, especially when those with power or status in the group overestimate their expertise.
3. *Average of members' opinions.* Group decision based on average of individual group members' opinions.
4. *Decision by authority after discussion.* Group in which designated leader makes decision after discussion with the group. Effectiveness often depends on the listening skills of the leader.
5. *Minority control.* Two or more members who constitute less than 50 percent of the group often makes decisions by (a) acting as an executive committee or (b) special problem solving sub group.

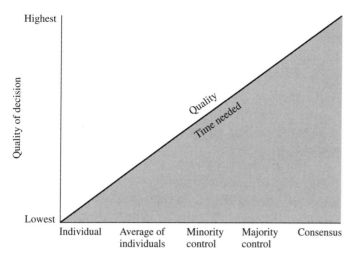

Figure 3.1 Decision type and quality

6. *Majority control.* Decision by a majority vote is the most common method used in the U.S. Discussion occurs only until at least 51 percent of the members decide on a course of action.
7. *Consensus.* Consensus is probably the most effective method of group decision making, but it also may take the most time. Perfect consensus is achieved when everyone agrees. A lesser degree of consensus is often accepted where everyone has had their say and will commit to the decision, but they may not completely agree with the decision.

They note that the quality of the decision and the time needed vary as a function of the number of people involved in the decision-making method, as shown in Figure 3.1.

David and Frank Johnson (1991) also list the following characteristics of effective decisions:

1. The resources of the group members are well used.
2. Time is well used.
3. The decision is correct, or of high quality.
4. The decision is put into effect fully by all the necessary members' commitments.
5. The problem-solving ability of the group is enhanced.

GROUP REFLECTION How well did your group do on each of these five characteristics of effective decisions?

Typically, novice decision-making groups don't take full advantage of the skills and talents of their members, and they often struggle to get started. Some researchers report a series of stages in team development (e.g., forming, storming, norming, performing) and offer suggestions for working through each stage (Scholtes, Joiner, and Streibel, 1996). Also, if you ask a group to invest time and effort in making a decision it is very important that the decision be implemented (or very good rationale provided for why it wasn't implemented). There are few things more frustrating than to be asked to spend lots of time and effort on work that goes nowhere.

CONFLICT MANAGEMENT

Conflict is a routine aspect of every project manager's job. *Conflict* is a situation in which an action of one person prevents, obstructs, or interferes with the actions of another person. On complex projects and tasks, highly talented and motivated people routinely disagree about the best ways to accomplish tasks and especially about how to deal with trade-offs among priorities. A conflict often is a moment of truth, since its resolution can follow either a constructive or a destructive path.

INDIVIDUAL REFLECTION Write the word *conflict* in the center of a blank piece of paper and draw a circle around it. Quickly jot down all the words and phrases you associate with the word *conflict* by arranging them around your circle.

Review your list of associations and categorize them as positive, negative, or neutral. Count the total number of positive, negative, and neutral associations, and calculate the percentage that are positive. Did you have more than 90 percent positive?

Less than 5 percent of the people who have done this Reflection in my classes and workshops have had more than 90 percent positive associations. The majority, in fact, have had less than 50 percent positive associations. Many have had less than 10 percent positive.

The predominance of negative associations with conflict is one of the reasons conflict management is so difficult for project managers. Many people prefer to avoid conflict or to suppress it when it does arise. They become fearful, anxious, angry, or frustrated; consequently, the conflict takes a destructive path.

The goal of this section is to help you develop a set of skills and procedures for guiding conflict along a more constructive path. I'd like to begin by asking you to complete a questionnaire to assess how you typically act in conflict situation. The "How I Act in Conflict" questionnaire is included as Exercise 2 at the end of this chapter. Take a few minutes to complete and score the questionnaire. Try to use professional conflicts and not personal conflicts as your point of reference.

Set the questionnaire aside for a few minutes and read Exercise 3, the Ralph Springer case study. Work through the exercise, completing the ranking form at the end.

GROUP ACTIVITY Share and discuss each member's results from Exercise 2. Discuss each of the possible ways to resolve the conflict.

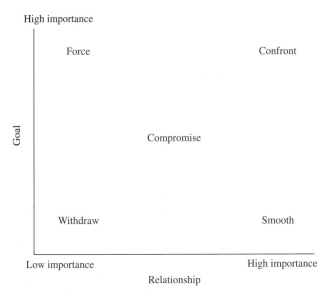

Figure 3.2 Blake and Mouton conflict model

Then compare your individual responses from Exercise 2 to your rankings in Exercise 3. Note that each of the alternatives listed in Exercise 3 represents one of the five strategies on the scoring form in Exercise 2. Match the alternatives to the strategies they represent. Discuss similarities and differences in the order each group member would have used the strategies and the relative effectiveness of each.

The five conflict strategies shown in Exercise 2—withdrawal, forcing, smoothing, compromise, and confrontation—were formulated into a model for analyzing approaches to conflict by Blake and Mouton (1964). The authors used two axes to represent the conflict strategies: (1) the importance of the goal, and (2) the importance of the task. The placement of each of the five strategies according to this framework is shown in Figure 3.2.

The five strategies are described as follows:

1. *Withdrawal.* Neither the goal nor the relationship is important—you withdraw from the interaction.
2. *Forcing.* The task is important but not the relationship—use all your energy to get the task done.
3. *Smoothing.* The relationship is more important than the task. You want to be liked and accepted.
4. *Compromise.* Both task and relationship are important, but there is a lack of time—you *both* gain and lose something.
5. *Confrontation.* Task and relationship are equally important. You define the conflict as a problem-solving situation and resolve through negotiation.

Each of these strategies is appropriate under certain conditions. For example, if neither the goal nor the relationship is important to you, then often the best thing to do is withdraw. If the relationship is extremely important and the task is not so important (at the time), then smoothing is appropriate. In many conflict situations, both the task and the relationship are important. In these situations, confronting and negotiating often lead to the best outcomes.

A *confrontation* is the direct expression of one opponent's view of the conflict, and his or her feelings about it, and an invitation to the other opponent to do the same.

Guidelines for Confrontation

1. Do not "hit and run." Confront only when there is time to jointly define the conflict and schedule a negotiating session.
2. Openly communicate your feelings about and perceptions of the issues involved in the conflict, and try to do so in minimally threatening ways.
3. Accurately and fully comprehend the opponent's views and feelings about the conflict.

Negotiation is a conflict resolution process by which people who want to come to an agreement try to work out a settlement.

Steps in Negotiating a Conflict

1. Confront the opposition.
2. Define the conflict mutually.
3. Communicate feelings and positions.
4. Communicate cooperative intentions.
5. Take the other person's perspective.
6. Coordinate the motivation to negotiate.
7. Reach an agreement that is satisfactory to both sides.

Constructively resolving conflicts through a confrontation–negotiation process takes time and practice to perfect, but it's worth it. Conflicts that do not get resolved at a personal level must be resolved at more time-consuming and costly levels—third-party mediation; arbitration; and, if all else fails, litigation.

Finally, here are some heuristics for dealing with conflicts in long-term personal and professional relationships:

1. Do not withdraw from or ignore the conflict.
2. Do not engage in "win–lose" negotiations.
3. Assess for smoothing.
4. Compromise when time is short.
5. Confront to begin problem-solving negotiations.
6. Use your sense of humor.

Remember that heuristics are reasonable and plausible, but not guaranteed. I suggest that you develop your own set of heuristics for dealing with conflict as well as for the other skills needed for effective teamwork. Some of my former students who now work as project managers emphasize during classroom visits that they spend a lot of time resolving conflicts—over meeting specifications, schedules, delivery dates, interpersonal differences among team members—and that most conflicts are dealt with informally.

Teamwork Challenges and Problems

REFLECTION What are some of the most common challenges and problems you've had working in groups? Please reflect for a moment. Make a list. Has a professor ever had you do this in your teams? If so, it's a clear indication that the professor understands the importance of group processing for identifying and solving problems.

What's on your list?

The challenges and problems you listed in the above reflection may have included the following:

- Members who don't show up for meetings or who don't show up prepared.
- Members who dominate the conversation.
- Members who don't participate in the conversation.
- Time wasted by off-task talk.
- Members who want to do the entire project because they don't trust others.
- Group meeting scheduling difficulties.
- No clear focus or goal.
- Lack of clear agenda, or hidden agendas.
- Subgroups excluding or ganging up on one or more members.
- Ineffective or inappropriate decisions and decision-making processes.
- Suppression of conflict or unpleasant flare-ups among group members.
- Members not doing their fair share of the work.
- Lack of commitment to the group's work by some members.

The problems listed above are commonly encountered by students (and professionals) working in groups. If they are not addressed they can turn a cooperative group into a pseudo group, as described in Chapter 2, where the group does worse than individuals working alone. If the challenges are addressed in a problem-solving manner, then the group is likely to perform at much higher levels (and the members will have a much more positive experience). The following process is widely used to address group problems.

Step 1: Identifying Challenges, Difficulties, and Barriers to Effective Group Work

- Reflect individually for a moment and start a list of challenges, barriers, or problems facing the group. Share the individual lists and create a joint list that includes at least one item from each group member.
- Do not solve (yet).
- Be realistic and specific.
- Work cooperatively.
- If more than one group is involved, list challenges, barriers, and problems for all groups on an overhead projector or flip chart.

Step 2: Addressing Barriers, Challenges, and Problems

1. Have each group or (if only one group is involved) each member select one item from the joint list.
2. Clarify: Make sure you have a common understanding of what the item means or represents.
3. Create three possible actions that will solve or eliminate the barrier.
4. Prioritize the possible solutions: Plan A, Plan B, Plan C.
5. Focus on what *will* work; be positive and constructive.
6. Implement the solutions; report back; celebrate and spread the ones that are effective.

Caveat: During implementation of group work expect some challenges, barriers, and problems. Doing so will help you recognize a roadblock when it appears. When it does appear, apply the appropriate parts of Step 2.

With one or more colleagues, develop three or more solutions. Implement one of these and then evaluate, replan, and retry.

The problem identification, problem formulation, and problem-solving format described above does not guarantee that your teamwork experiences will be free from troubles. But having a format for getting problems out on the table and then dealing with them in a problem-solving manner usually reduces the frustration and interference of group problems.

REFLECTION: TEAMWORK

I've tried to address many of the highlights of effective teamwork and team problem solving, but I've barely scratched the surface. Hundreds of books and articles have been written on effective teamwork, and I've listed a few of my favorites in the reference section (see, e.g., Fisher, Rayner, and Belgard, 1995; Goldberg, 1995; Hackman, 1990; Katzenbach and Smith, 1993*a,* 1993*b*). As I mentioned earlier, one of the most widely used teamwork books is Scholtes, Joiner, and Streibel's (1996) *The Team Handbook.*

QUESTIONS

1. What other skills besides those mentioned in this chapter do you feel are essential for successful groups? How about trust and loyalty, for example? I briefly dealt with trust and loyalty under the organization section, but you may want to emphasize them more. Check the references (e.g., David Johnson and Frank Johnson, 1991) for more. What other teamwork skills would you like to follow up on?
2. What are some of the strategies for developing a good set of working conditions in a group?
3. What are your reactions to the list of characteristics of admired leaders in Table 3.3? Were you surprised by the high ranking of honesty?
4. Why is conflict central to effective teamwork and project work? What are some strategies for effectively managing conflict?
5. Keep a log of problems you've faced in working on project teams. How do the problems change over the life of the group?
6. The next time a problem occurs in a group, try the problem-solving process outlined in the chapter. How well did it work?

EXERCISES

1. THEY'LL NEVER TAKE US ALIVE

Product or Activity	Ranking	Number of Fatalities
Accidents		
Alzheimer's disease		
Blood poisoning		
Cancer		
Diabetes		
Hardening of arteries		
Heart disease		
HIV and AIDS		
Homicide		
Kidney disease		
Liver disease		
Lung disease		
Pneumonia and influenza		
Stroke		
Suicide		

SOURCES: Office of the Surgeon General; National Center for Health Statistics.

On the accompanying chart, in alphabetical order, are listed the top 15 causes of death in the United States in 1997. The data were taken from an annual review of death certificates. Your task is to rank the products and activities in order of the number of deaths they cause each year. Place the number 1 next to the one that causes the most deaths, the number 2 by the one that causes the second most deaths, and so forth. Then, write in your estimate of the number of fatalities each product or activity causes.

Group Tasks

1. After individuals have filled in the chart, determine one ranking for the group. (Do not worry yet about the estimates for the number of fatalities.)
2. Every group member must be able to explain the rationale for the group's ranking.
3. When your group finishes, and each member has signed the chart, (*a*) record your estimated number of fatalities in the U.S. for each, and then (*b*) compare your rankings and estimates with those of another group.

2. How I Act in Conflict

The proverbs listed in the accompanying table can be thought of as descriptions of some of the different strategies for resolving conflicts. Proverbs state conventional wisdom, and the ones listed here reflect traditional wisdom for resolving conflicts. Read each carefully. Using the scale provided, indicate how typical each proverb is of your actions in a conflict. Then score your responses on the chart at the end of the table. The higher the total score in each conflict strategy, the more frequently you tend to use that strategy. The lower the total score for each conflict strategy, the less frequently you tend to use that strategy.

5 = Very typical of the way I act in a conflict
4 = Frequently typical of the way I act in a conflict
3 = Sometimes typical of the way I act in a conflict
2 = Seldom typical of the way I act in a conflict
1 = Never typical of the way I act in a conflict

_____ 1. It is easier to refrain than to retreat from a quarrel.
_____ 2. If you cannot make a person think as you do, make him or her do as you think.
_____ 3. Soft words win hard hearts.
_____ 4. You scratch my back, I'll scratch yours.
_____ 5. Come now and let us reason together.
_____ 6. When two quarrel, the person who keeps silent first is the most praiseworthy.
_____ 7. Might overcomes right.
_____ 8. Smooth words make smooth ways.
_____ 9. Better half a loaf than no bread at all.
_____ 10. Truth lies in knowledge, not in majority opinion.

_____ 11. He who fights and runs away lives to fight another day.
_____ 12. He hath conquered well that hath made his enemies flee.
_____ 13. Kill your enemies with kindness.
_____ 14. A fair exchange brings no quarrel.
_____ 15. No person has the final answer, but every person has a piece to contribute.
_____ 16. Stay away from people who disagree with you.
_____ 17. Fields are won by those who believe in winning.
_____ 18. Kind words are worth much and cost little.
_____ 19. Tit for tat is fair play.
_____ 20. Only the person who is willing to give up his or her monopoly on truth can ever profit from the truths that others hold.
_____ 21. Avoid quarrelsome people, for they will only make your life miserable.
_____ 22. A person who will not flee will make others flee.
_____ 23. Soft words ensure harmony.
_____ 24. One gift for another makes good friends.
_____ 25. Bring your conflicts into the open and face them directly; only then will the best solution be discovered.
_____ 26. The best way of handling conflicts is to avoid them.
_____ 27. Put your foot down where you mean to stand.
_____ 28. Gentleness will triumph over anger.
_____ 29. Getting part of what you want is better than not getting anything at all.
_____ 30. Frankness, honesty, and trust will move mountains.
_____ 31. There is nothing so important that you have to fight for it.
_____ 32. There are two kinds of people in the world, the winners and the losers.
_____ 33. When someone hits you with a stone, hit him or her with a piece of cotton.
_____ 34. When both people give in halfway, a fair settlement is achieved.
_____ 35. By digging and digging, the truth is discovered.

Scoring

Withdrawal	Forcing	Smoothing	Compromise	Confrontation
1.	2.	3.	4.	5.
6.	7.	8.	9.	10.
11.	12.	13.	14.	15.
16.	17.	18.	19.	20.
21.	22.	23.	24.	25.
26.	27.	28.	29.	30.
31.	32.	33.	34.	35.
Total	Total	Total	Total	Total

SOURCE: David Johnson and Roger Johnson, 1991.

3. Case Study—Ralph Springer

The following case gives you a chance to apply the Blake and Mouton (1964) conflict model to a hypothetical situation. Read the case carefully and then label each of the possible actions from most to least effective and from most to least likely.

> You have been working as a project manager in a large company for some time. You are friends with most of the other project managers and, you think, respected by all of them. A couple of months earlier, Ralph Springer was hired as a supervisor. He is getting to know the other project managers and you. One of the project managers in the company, who is a friend of yours, confided in you that Ralph has been saying rather nasty things about your looks, the way you dress, and your personal character. For some reason you do not understand, Ralph has taken a dislike to you. He seems to be trying to get other project managers to dislike you also. From what you hear, there is nothing too nasty for him to say about you. You are worried that some people might be influenced by him and that some of your co–project managers are also beginning to talk about you behind your back. You are terribly upset and angry at Ralph. Since you have a good job record and are quite skilled in project management, it would be rather easy for you to get another job.

Rank each of the following five courses of action from 1 (most effective, most likely) to 5 (least effective, least likely). Use each number only once. Be realistic.

Effective	Likely	
_____	_____	I lay it on the line. I tell Ralph I am fed up with the gossip. I tell him that he'd better stop talking about me behind my back, because I won't stand for it. Whether he likes it or not, he is going to keep his mouth shut about me or else he'll regret it.
_____	_____	I try to bargain with him. I tell him that if he will stop gossiping about me I will help him get started and include him in the things other project managers and I do together. I tell him that others are angry about the gossiping and that it is in his best interest to stop. I try to persuade him to stop gossiping in return for something I can do.
_____	_____	I try to avoid Ralph. I am silent whenever we are together. I show a lack of interest whenever we speak, look over his shoulder and get away as soon as possible. I want nothing to do with him for now. I try to cool down and ignore the whole thing. I intend to avoid him completely if possible.
_____	_____	I call attention to the conflict between us. I describe how I see his actions and how it makes me feel. I try to begin a discussion in which we can look for a way for him to stop making me the target of his conversation and a way to deal with my anger. I try to see things from his viewpoint and seek a solution that will suit us both. I ask him how he feels about my giving him this feedback and what his point of view is.
_____	_____	I bite my tongue and keep my feelings to myself. I hope he will find out that the behavior is wrong without my saying anything. I try to be extra nice and show him that he's off base. I hide my anger. If I tried to tell him how I feel, it would only make things worse.

References

Blake, R. R., and J. S. Mouton. 1964. *The managerial grid.* Houston: Gulf Publishing Company.

Fisher, Kimball, Steven Rayner, and William Belgard. 1995. *Tips for teams: A ready reference for solving common team problems.* New York: McGraw-Hill.

Goldberg, David E. 1995. *Life skills and leadership for engineers.* New York: McGraw-Hill.

Hackman, J. R. 1990. *Groups that work (and those that don't): Creating conditions for effective teamwork.* San Francisco: Jossey-Bass.

Johnson, David W., and Frank P. Johnson, 1991. *Joining together: Group theory and group skills,* 4th ed. Englewood Cliffs, NJ: Prentice-Hall.

Johnson, David W., and Roger T. Johnson, 1991. *Teaching students to be peacemakers.* Edina, MN: Interaction Book Company.

Katzenbach, Jon R., and Douglas K. Smith. 1993*a. The wisdom of teams: Creating the high-performance organization.* Cambridge, MA: Harvard Business School Press.

———. 1993*b.* The discipline of teams. *Harvard Business Review* 71(2): 111–20.

Kouzes, J. M., and B. Z. Posner. 1987. *The leadership challenge: How to get extraordinary things done in organizations.* San Francisco: Jossey-Bass.

———. 1993. *Credibility: How leaders gain and lose it, why people demand it.* San Francisco: Jossey-Bass.

McNeill, Barry, Lynn Bellamy, and Sallie Foster. 1995. *Introduction to engineering design.* Tempe, Arizona: Arizona State University.

Napier, Rodney W., and Matti K. Gershenfeld. 1973. *Groups: Theory and experience.* Boston: Houghton Mifflin.

Scholtes, Peter R. 1998. *The leader's handbook: Making things happen, getting things done.* New York: McGraw-Hill.

Scholtes, Peter R., Brian L. Joiner, and Barbara J. Streibel. 1996. *The team handbook,* 2nd ed. Madison, WI: Joiner Associates.

Taylor, James. 1999. *A survival guide for project managers.* New York: AMACOM.

chapter 4

PROJECT MANAGEMENT PRINCIPLES AND PRACTICES

This chapter discusses what a project is and explains why projects and project management are receiving a lot of attention right now. The number of books and articles on project management is growing almost exponentially. Something is happening here. Perhaps it is due in part to observations like those in a recent Project Management Institute survey, which indicated that only a fraction of technology projects in the United States finish on time. The percentages are startling: Close to half of the projects started were never finished, and 30 percent were completed but took at least twice as long as expected; some took five times as long. Only 10 percent of the projects were finished on time.

The situation has changed a lot since the development of scheduling tools and strategies such as the critical path method (CPM) and the program evaluation and review technique PERT) in the 1950s. Laufer, Denker, and Shenhar (1993) have outlined the evolution in the nature of project management. A summary of the changes is shown in Table 4.1. Laufer et al. emphasize that even as projects have become more complex, the time to accomplish them has become shorter; thus, many projects require simultaneous management.

Ed Yourdon claims in his book *Death March Projects* (1997) that many projects must be completed in half the time, with half the budget, or with half the resources initially planned, hence the phrase "death march projects." However, Yourdon also claims that it is possible, almost exciting at times, to be a part of this type of project.

REFLECTION   Think about your involvement with projects both in school and in other aspects of your life. Have you been involved in more and more projects in school? Think about some of these projects. What are the distinguishing features of the projects you've been involved with? For example, did you have to make a presentation or write a report? Did you

Table 4.1 Evolution of models of project management

Central Concept	Era of Model	Dominant Project Characteristics	Main Thrust	Metaphor	Means
Scheduling (control)	1960s	Simple, certain	Coordinating activities	Scheduling regional flights in an airline	Information technology, planning specialists
Teamwork (integration)	1970s	Complex, certain	Cooperation between participants	Conducting a symphony orchestra	Process facilitation, definition or roles
Reducing uncertainty (flexibility)	1980s	Complex, uncertain	Making stable decisions	Exploring an unknown country	Search for information, selective redundancy
Simultaneity (dynamism)	1990s	Complex, uncertain, quick	Orchestrating contending demands	Directing a three-ring circus with continuous program modification based on live audience feedback	Experience, responsiveness, and adaptability

have to give a performance, as in the production of a dramatic event such as a play? Have you participated in a science fair project or a design project? Please take a few minutes to reflect on changes in your involvement in projects, and make a list of the distinguishing features of these projects.

What's on your list? Does it include items such as "common, overriding purpose and established goals"; "temporary" (i.e., clear beginning and end); "one-time activity"; "requires coordinating many interrelated activities"; or "involves several people"?

WHAT IS A PROJECT?

A dictionary of project management terms (Cleland and Kerzner, 1985) defines *project* as follows:

> [A project is] a combination of human and nonhuman resources pulled together in a temporary organization to achieve a specified purpose.

Project is defined by Snead and Wycoff (1997) as "a nonroutine series of tasks directed toward a goal." In their helpful guide, the authors claim that "success depends on the ability to effectively complete projects."

A textbook (Nicholas, 1990) that I have used in my project management classes lists the following features of projects:

- Definable purpose with established goals.
- Cost, schedule, and performance requirements.
- Multiple resources across organizational lines.
- One-time activity.
- Element of risk.
- Temporary activity.
- Process of phases; project life cycle.

Based on this list of features you can see that projects are quite different from the ongoing, day-to-day work that most of us do. Each project is unique, is temporary, has an element of risk, and has a definable purpose with established goals. Two features of projects that I'd like to explore further are (1) cost, schedule, and performance requirements, and (2) process phases or project life cycle.

KEYS TO PROJECT SUCCESS

Traditionally, project success has been measured according to three criteria: cost, time, and performance. Although students in classes often negotiate time (especially due dates) and performance requirements, there is often less flexibility in professional life. For example, the due dates for submitting research proposals to funding agencies are rigid. One must get the proposal in before the deadline or wait until next year (and hope the agency still is making grants in that particular area). In many large construction companies there are significant incentives for finishing a project on time, and major penalties for finishing late. Some projects have been terminated when there were cost overruns; note the tragic demise of the Superconducting Supercollider (the multibillion dollar particle accelerator in Texas that was terminated by the U.S. Congress).

Subsequent chapters of this book will explore how cost, time, and performance are operationalized, that is, how they are put into practice. Briefly, cost is operationalized by budgets, time by schedules, and performance by specifications.

Cost, time, and performance. Is this it? Is this all that we need to attend to for successful projects? Many project management experts are discussing a fourth aspect of project success—client acceptance. Pinto and Kharbanda (1995), for example, maintain that there is a quadruple constraint on project success, which of course, increases the challenge of completing projects successfully (see Figure 4.1).

The most common way in which client acceptance is operationalized is to involve the client throughout the project. One of the most famous examples of this is Boeing's 777 project, in which customers were involved early on and throughout the project. These customer airlines had a significant influence on how the 777 was designed and built. Boeing's vision was to build a high-quality aircraft in an environment of no secrecy and no rivalry. These new values were clarified in the following three statements (cited in Snead and Wycoff, 1997):

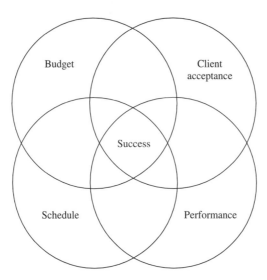

Figure 4.1 Project success: quadruple constraint

SOURCE: Pinto and Khorbanda, 1995.

1. Use a style of management unheard of in the industry: working together while building trust, honesty, and integrity.
2. In the past, people were afraid to state a problem because of the practice of killing the messenger. We will instead celebrate our problems and get them out into the open so we can work on them.
3. We must come with no limitations in our mind. We must have a shared thought, vision, appreciation, and understanding of what we are going to accomplish together.

Boeing's long-range goals for the 777 helped create the environment described above:

- Design, develop, and produce a plane safer and more reliable than any other plane in aviation history that is state-of-the-art and service-ready on delivery, to be called the 777.
- Design, develop, and produce a program to empower a massive team of people to implement the "working together" philosophy while creating the 777.

Phil Condit, Boeing's CEO, said, "The task for us at Boeing is to provide a massive change in thinking throughout the company—this is a cultural shift, and it isn't easy!" Boeing experienced many positive changes (and outcomes) during this process. The 777 was delivered on time and was under budget. Most significantly, however, it positively changed the "management-teamwork" paradigm from a hierarchical relationship to a lateral relationship.

If you'd like to explore Boeing's 777 project in more detail, the book *21st Cen-*

tury Jet, by Karl Sabbagh, and the six-part PBS video series based on the book provide rich insight into the process.

PROJECT LIFE CYCLE

REFLECTION   Please reconsider the projects that came to mind during the Reflection at the beginning of this chapter. Did each project seem to go through a series of stages? If so, how would you characterize them? Think about how the activities and work on the project changed from beginning to end. Jot down your reflections.

The prevailing view of the project life cycle is that projects go through distinct phases, such as the following:

- Conceiving and defining the project
- Planning the project
- Implementing the plan
- Completing and evaluating the project
- Operating and maintaining the project

A typical construction project has the following seven phases (Kerzner, 1998):

1. Planning, data gathering, and procedures
2. Studies and basic engineering
3. Major review
4. Detail engineering
5. Detail engineering/construction overlap
6. Construction
7. Testing and commissioning

Some people, however, perhaps in moments of frustration, have described the phases of a project in a more cynical way:

1. Wild enthusiasm
2. Disillusionment
3. Total confusion
4. Search for the guilty
5. Punishment of the innocent
6. Praise and honors for the nonparticipants

These faults could often be avoided if project managers think about resource distribution over the project life cycle.

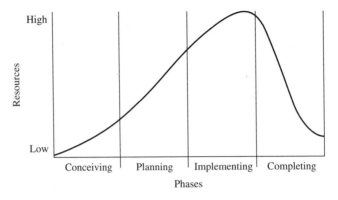

Figure 4.2 Resource distribution over the project life cycle

REFLECTION   Consider the first four phases of the project life cycle described above (conceiving, planning, implementing, and completing) and sketch how you think resources (people, money, etc) are distributed throughout the life of a project. What did you come up with? Continually increasing resources? Increasing then decreasing? Why did you draw the shape you did?

A common distribution of resources (people, materials, etc.) is shown in Figure 4.2.

Project managers must also consider how their ability to make changes and the cost of those changes vary over the project life cycle. Figure 4.3 shows the relationship between these two factors. Consider the essential message in Figure 4.3. Since

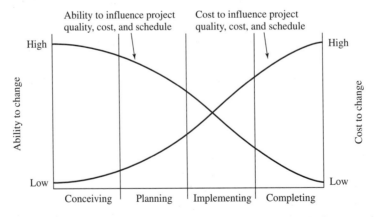

Figure 4.3 Ability to change and cost to make changes over the project life cycle

you have considerably more flexibility early in a project and it's cheaper to make changes, don't skimp on planning during the early stages. Though this essential message probably makes a lot of sense, it's one that is hard to implement. Because of the extraordinary eagerness of many project managers and project team members to get going, careful and thorough planning is often neglected. This essential message could also be described as a project management heuristic. (See Chapter 1 for elaboration on the meaning of *heuristics* and their importance in engineering.) Additional examples of project management heuristics are the following:

- Allocate resources to the weak link.
- Freeze the design—at some stage in the project (when about 75 percent of the time or resources are used up), the design must be frozen.
- Periodically discuss the process and ask meta-level questions (e.g., What are we doing? Why are we doing it? How does it help?).

A superb collection of modeling heuristics highly relevant to project management was presented by Ravindran, Phillips, and Solberg (1987):

1. Do not build a complicated model when a simple one will suffice.
2. Beware of molding the problem to fit the technique.
3. The deduction phase of modeling must be conducted rigorously.
4. Models should be validated prior to implementation.
5. A model should never be taken too literally.
6. A model should neither be pressed to do, nor criticized for failing to do, that for which it was never intended.
7. Beware of overselling a model.
8. Some of the primary benefits of modeling are associated with the process of developing the model.
9. A model cannot be any better than the information that goes into it.
10. Models cannot replace decision makers.

The heuristics given in both of the above lists are important when thinking about the project life cycle and will become crucially important when we look at the use of project scheduling models later in this book.

PROJECT PLANNING

Projects typically start with a statement of work (SOW) provided by the client. The statement of work is a narrative description of the work required for the project. In engineering classes, the statement of work is provided by the faculty member. Planning starts with the development of a work breakdown structure (WBS). A WBS is "a deliverable-oriented grouping of project elements which organizes and defines the total scope of a project" (Duncan, 1996). There are typically three to six levels in a

WBS, such as program, project, task, and subtask. Developing a work breakdown structure is important for "scoping" a project, that is, determining the specific tasks that have to be completed, choosing appropriate groupings for these activities, and setting precedence and interdependence (what has to follow what and what can be going on at the same time).

These two parts of project planning—the statement of work and the work breakdown structure—are often neglected in traditional project management textbooks and classes, perhaps due to the eagerness to get to the nitty-gritty of project scheduling by critical path analysis. However, carefully considering these two initial aspects of project scoping is an important part of not skimping on planning.

REFLECTION: PROJECT MANAGEMENT

Project management is a relatively new profession and is growing at a remarkable rate. *Fortune* magazine called project management "Career Number 1" for the 1990s. When I was in engineering school in the late 1960s, project management courses weren't offered. Since I now teach several project management courses, I've had to learn it through experience and research. Several of the books I've found useful and have used as texts in my courses are listed in the references (Culp and Smith, 1992; Eisner, 1997; Frame, 1994, 1995; Grady, 1992; Graham and Englund, 1997; Kerzner, 1998; Lewis, 1993, 1995a, 1995b, 1998; Lientz and Rea, 1995; Meredith and Mantel, 1994; Nicholas, 1990; Pinto and Kharbanda, 1995; Snead and Wycoff, 1997; Taylor, 1999). Since project management is an emerging field and is changing quite rapidly, I encourage you to continue honing your skills and competencies.

QUESTIONS

1. What is a project? What are the key characteristics of projects? How does project management differ from management in general?
2. What are the three cardinal conditions of project success?
3. What has been your experience in completing projects on time, under budget, within specifications, and acceptable to the client? What is your batting average? Is it better than the 10 percent figure cited by the Project Management Institute survey?
4. How does your distribution of effort vary over the life of projects that you've worked on? Do you start strong and taper off? Or do you start slowly and build? Sketch out a graph that has effort on the ordinate (y) and time on the abscissa (x) for a typical project. Is your effort curve consistent with the bell-shaped curve shown in Figure 4.2? Is it different? Does most of your effort go in to the last few hours before the project is due? If much of your effort is applied in the closing hours of the project, perhaps you are "freezing the design" too late. How does your enthusiasm vary over the project life cycle?

5. Start developing a list of your own project management heuristics. There are several books that list rules of thumb, which are one type of heuristic. One of my favorites is by Grady (1992).

EXERCISE

PROJECT PLANNING

Now that you've had an opportunity to think about projects, project management, project life cycles, and project scoping, I'd like you to try applying what you've learned.

Suppose you have two tickets to a fabulous concert and are planning a special dinner for two prior to the concert. Your menu consists of a very special soup and baked chicken entrée. The soup must be boiled for 35 minutes, and you should allow 15 minutes to serve and consume it. The chicken dish requires a fair amount of preparation: you have to boil the rice for 30 minutes, brown the chicken in the frying pan for 15 minutes, and place the rice and chicken in a baking dish in the oven for 15 minutes. It takes 5 minutes to prepare a sauce in the frying pan and 15 minutes to boil the peas. (You only have two pots and one frying pan.) You have bought a good red wine; allow 5 minutes to uncork it (very carefully) and 30 minutes to let it stand before serving it. You plan to allow 25 minutes to serve and consume the entrée and wine. How much time do you need to prepare and consume the meal?

What representation (model) did your group use to determine the time? How did you keep track of which activities had to follow others and which could be going on at the same time?

REFERENCES

Cleland, D. I., and H. Kerzner. 1985. *A project management dictionary of terms.* New York: Van Nostrand Reinhold.

Culp, G., and A. Smith. 1992. *Managing people (including yourself) for project success.* New York: Van Nostrand Reinhold.

Duncan, William R. 1996. *A guide to the project management body of knowledge.* Newton Square, PA: Project Management Institute.

Eisner, H. 1997. *Essentials of project management and systems engineering management.* New York: Wiley.

Frame, J. D. 1994. *The new project management.* San Francisco: Jossey-Bass.

———. 1995. *Managing projects in organizations.* San Francisco: Jossey-Bass.

Grady, Robert B. 1992. *Practical software metrics for project management and process improvement.* Englewood Cliffs, NJ: Prentice-Hall.

Graham, Robert J., and Randall L. Englund. 1997. *Creating an environment for successful projects.* San Francisco: Jossey-Bass.

Kerzner, H. 1998. *Project management: A systems approach to planning, scheduling, and controlling,* 6th ed. New York: Van Nostrand Reinhold.

Laufer, A., G. R. Denker, and A. J. Shenhar. 1993. Simultaneous management: The key to excellence in capital projects. Unpublished manuscript.

Lewis, James P. 1993. *The project manger's desk reference: A comprehensive guide to project planning, scheduling, evaluation, control & systems.* New York: Probus.

———. 1995a. *Project planning, scheduling, & control: A hands-on guide to bringing projects in on time and on budget.* New York: Probus.

———. 1995b. *Fundamentals of project management.* New York: AMACOM.

———. 1998. *Mastering project management: Applying advanced concepts of systems thinking, control and evaluation, resource allocation.* New York: McGraw-Hill.

Lientz, Bennet, and Kathryn Rea. 1995. *Project management for the 21st century.* San Diego: Academic Press.

Meredith, J.R., and S. J. Mantel. 1994. *Project management: A managerial approach.* New York: Wiley.

Nicholas, J. M. 1990. *Managing business & engineering projects: Concepts and implementation.* Englewood Cliffs, NJ: Prentice-Hall.

Pinto, J. K., and O. P. Kharbanda. 1995. *Successful project managers: Leading your team to success.* New York: Van Nostrand Reinhold.

Ravindran, A., D. T. Phillips, and J. J. Solberg. 1987. *Operations research: Principles and practices.* New York: Wiley.

Sabbagh, Karl. 1997. *21st-century jet: The making and marketing of the Boeing 777.* New York: Scribner.

Snead, G. L., and J. Wycoff. 1997. *To do, doing, done! A creative approach to managing projects and effectively finishing what matters most.* New York: Fireside.

Taylor, James. 1999. A survival guide for project managers. New York: AMACOM.

Yourdon, Ed. 1997. *Death march projects.* Reading, MA: Addison-Wesley.

chapter

5

PROJECT MANAGER'S ROLE

Project management is undergoing enormous changes, as Table 4.1 indicated, and thus the role of the project manager is changing. Before we explore the changes that are occurring in project management, let's explore changes that are occurring in business, industry, government, and education.

INDIVIDUAL AND GROUP REFLECTION Think about changes that have occurred in the workplace (or school if that is your principal area of experience) in the past five years. Make a list of some of the most notable changes and compare it with other team members' lists.

Students in my project management classes who do the above Reflection come up with lots of changes they're noticing—communications technology, computers, global marketplace, emphasis on quality, shortened time frames, and changing role and importance of knowledge workers.

CHANGES IN THE WORKPLACE

Changes in the workplace have been studied and summarized by numerous authors, including Byrne (1992). Changes occurring in how engineers work in business and industry, summarized in Table 5.1, have serious implications for how we prepare engineering graduates for working in the 21st century.

The changes that are occurring in business and industry suggest that we should consider changes in engineering education to prepare our graduates to function effectively in the "new paradigm" companies. The "Made in America" study

Table 5.1 Reinventing the corporation

	The Current Model	21st-Century Prototype
Organization	Hierarchy	Network
Structure	Self-sufficiency	Interdependencies
Worker expectations	Security	Personal growth
Leadership	Autocratic	Inspirational
Workforce	Homogeneous	Culturally diverse
Work	By individuals	By teams
Markets	Domestic	Global
Advantage	Cost	Time
Focus	Profits	Customers
Resources	Capital	Information
Governance	Board of directors	Varied constituents
Quality	What's affordable	No compromises

SOURCE: Byrne, 1992.

(Dertouzos, Lester, and Solow, 1989) recommended that the Massachusetts Institute of Technology should

1. Broaden its educational approach in the sciences, in technology, and in the humanities and should educate students to be more sensitive to productivity, to practical problems, to teamwork, and to the cultures, institutions, and business practices of other countries.

2. Create a new cadre of students and faculty characterized by (1) interest in, and knowledge of, real problems and their societal, economic, and political context; (2) an ability to function effectively as members of a team creating new products, processes, and systems; (3) an ability to operate effectively beyond the confines of a single discipline; and (4) an integration of a deep understanding of science and technology with practical knowledge, a hands-on orientation, and experimental skills and insight.

3. Revise subjects to include team projects, practical problems, and exposure to international cultures. Encourage student teaching to instill a stronger appreciation of lifelong learning and the teaching of others. Reinstitute a foreign-language requirement in the undergraduate admissions process.

Changes in engineering education were described in a paper in the Frontiers in Education Conference proceedings (Smith and Waller, 1997) and summarized in Table 5.2. If you're interested in learning more about new paradigms for engineering education, you may view the paper on the World Wide Web at the ASEE/IEEE Frontiers in Education page.

The premier issue of an exciting new engineering magazine, *Today's Engineer,* proposes that we are at the dawning of a new age of engineering—the crossroads for

Table 5.2 Comparison of old and new paradigms for college teaching

	Old Paradigm	New Paradigm
Knowledge	Transferred from faculty to students	Jointly constructed by students and faculty
Students	Passive vessel to be filled by faculty's knowledge	Active constructor, discoverer, transformer of knowledge
Mode of learning	Memorizing	Relating
Faculty purpose	Classify and sort students	Develop students' competencies and talents
Student goals	Complete requirements, achieve certification within a discipline	Grow, focus on continual lifelong learning within a broader system
Relationships	Impersonal relationship among students and between faculty and students	Personal transaction among students and between faculty and students
Context	Competitive/individualistic	Cooperative learning in classroom and cooperative teams among faculty
Climate	Conformity/cultural uniformity	Diversity and personal esteem; cultural diversity and commonality
Power	Faculty holds and exercises power, authority, and control	Students are empowered; power is shared among students and between students and faculty
Assessment	Norm-referenced (i.e., graded "on the curve"); typically multiple choice items; student rating of instruction at end of course	Criterion-referenced; typically performances and portfolios; continual assessment of instruction
Ways of knowing	Logico-scientific	Narrative
Epistemology	Reductionist; facts and memorization	Constructivist; inquiry and invention
Technology use	Drill and practice; textbook substitute; chalk and talk substitute	Problem solving, communication, collaboration, information access, expression
Teaching assumption	Any expert can teach	Teaching is complex and requires considerable training

SOURCE: Smith and Waller, 1997.

a changing professional model (Gaynor, 1998). The editor, Gaynor, claims that this new model makes three demands on us: that we transcend traditional boundaries, that we think strategically, and that we develop a business perspective. Gaynor also says that technical competence is an absolute requirement, but by itself is no longer sufficient. It must be integrated with breadth of vision, flexibility, customer focus, and business orientation.

Changes like those outlined by Dertouzos and his colleagues (1989), as well as Gaynor (1998), are enormously difficult to implement in a direct, linear manner. The nature of change is described by Katzenbach and Smith (1993) through a "whitewater raft ride" metaphor. The authors also list behavioral changes that are demanded by change.

> Major change, by its nature, is intentionally disruptive and largely unprogrammable. In comparing the management of major versus normal change, one top executive said, "It used to be like I-75. You'd lay it out from Toledo to Tampa. Now it's more like a whitewater raft ride. You try to get the right people in the raft and do the best you can to steer it. But you never know what's just around the bend." (p. 208)

Katzenbach and Smith suggest several behavioral changes that will help us perform in today's workplace (see Table 5.3).

Table 5.3 Behavioral changes demanded by performance in the 1990s and beyond

From	To
Individual accountability	Mutual support, joint accountability, and trust-based relationships *in addition* to individual accountability
Dividing those who think and decide from those who work and do	Expecting everyone to think, work, and do
Building functional excellence through each person executing a narrow set of tasks ever more efficiently	Encouraging people to play multiple roles and work together interchangeably on continuous improvement
Relying on managerial control	Getting people to buy into meaningful purpose, to help shape direction, and to learn
A fair day's pay for a fair day's work	Aspiring to personal growth that expands as well as exploits each person's capabilities

Peter Drucker (1993), who has written more articles for the *Harvard Business Review* than anyone else, recently described the changing views of the "manager" concept. Drucker stresses the idea of the "knowledge worker" and, consistent with this concept, focuses on skills and strategies for "managing the knowledge worker." In the 1920s, a manager was seen as one who was responsible for the work of subordinates; in the 1950s, a manager was one who was responsible for the performance of people; and in the 1990s, and beyond, a manager is one who is responsible for the application and performance of knowledge.

INDIVIDUAL REFLECTION How are you feeling about all these impending changes? A bit overwhelmed, no doubt. Are you seeing changes in your educational experience? Is your college education on the cutting edge of modern practice? What do you think about the state of your education? Discuss with your group.

Changes in Project Management

Since so many changes are occurring in the workplace—including downsizing, rightsizing, and attending to the customer—is there any question that change is also occurring in project management? Management guru Tom Peters (1999) makes bold claims about the importance of project management in the following statement: "Those organizations that take project management seriously as a discipline, as a way of life, are likely to make it into the 21st century. Those that do not are likely to find themselves in good company with dinosaurs." (p. 128) Peters (1991), an engineering graduate who wrote a masters thesis on PERT charts, has also made the following statements:

- "Tomorrow's corporation is a 'collection of projects.'"
- "Everyone needs to learn to work in teams with multiple independent experts—each will be dependent upon all the others voluntarily giving their best."
- "The new lead actor/boss—the Project Manager—must learn to command and coach; that is, to deal with paradox." (p. 64)

In the area of project management, several authors have summarized the most notable changes. Pinto and Kharbanda (1995) refer to our age as "The Age of Project Management." Key features of this age are the following:

1. Shortened market windows and product life cycles.
2. Rapid development of third-world and closed economies.
3. Increasingly complex and technical products.
4. Heightened international competition.
5. The environment of organizational resource scarcity.

Lientz and Rea (1995) list several trends that affect projects:

Global competition	Empowerment
Rapid technological change	Focus on quality and continuous improvement
Product obsolescence	Measurement
Organizational downsizing	Interorganizational systems
Business reengineering	

Furthermore, Lientz and Rea remind us that projects are set in time. They are also set in the context of organization, a legal system, a political system, a technology structure, an economic system, and a social system. These environmental factors do affect projects, and the project manager must respond to the resulting challenges.

If I haven't yet convinced you that there are many changes occurring in the business world and that the emergence of project management is one of them, try the following Reflection.

INDIVIDUAL REFLECTION What does it take to be a good project manager? Take a few minutes to think about the skills and competencies (and perhaps the attitudes) needed for effective project management. Make a list. Compare your list with those of other students.

Do you know any project managers? Do you have relatives or friends who do project work? Try to find someone you can interview to help you get your bearings on project management. (See the exercise at the end of this chapter.) Then revise your list.

Skills Necessary for Effective Project Managers

Barry Posner (1987) conducted a survey of project managers, asking them what it takes to be a good project manager. He got the following results:

1. Communications (84 percent of the respondents listed it)
    a. Listening
    b. Persuading
2. Organizational skills (75 percent)
    a. Planning
    b. Goal setting
    c. Analyzing
3. Team-building skills (72 percent)
    a. Empathy
    b. Motivation
    c. Esprit de corps
4. Leadership skills (68 percent)
    a. Sets example
    b. Energetic
    c. Vision (big picture)
    d. Delegates
    e. Positive
5. Coping skills (59 percent)
    a. Flexibility
    b. Creativity
    c. Patience
    d. Persistence
6. Technological skills (46 percent)
    a. Experience
    b. Project knowledge

Several authors have surveyed project managers and conducted extensive literature searches to learn about essential project management skills. Pinto and Kharbanda (1995) list the following skills necessary for effective project managers:

- *Planning:* work breakdown, project scheduling, knowledge of project management software, budgeting and costing.
- *Organizing:* team building, establishing team structure and reporting assignments, defining team policies, rules, and protocols.
- *Leading:* motivation, conflict management, interpersonal skills, appreciation of team members' strengths and weaknesses, reward systems.
- *Controlling:* project review techniques, meeting skills, project close-out techniques.

Lientz and Rea (1996) provide the following list of keys to success as a project manager:

Communicate regularly *in person* with key team members.

Keep management informed.

Keep informed on all aspects of the project.

Delegate tasks to team members.

Listen to input from team members.

Be able to take criticism.

Respond to and/or act on suggestions for improvement.

Develop contingency plans.

Address problems.

Make decisions.

Learn from past experience.

Run an effective meeting.

Set up and manage the project file.

Use project management tools to generate reports.

Understand trade-offs involving schedule and budget.

Have a sense of humor.

INDIVIDUAL REFLECTION   How do these lists compare with yours? Was there lots of overlap? Were there categories of items that were on your list but not on these, and vice versa?

Research by Jeffrey Pinto (1986) sought to quantify some of these factors by correlating them with their importance for system implementation (see the accompanying box). "System implementation" may be interpreted as a successful project outcome.

How does one implement all of the characteristics of effective project managers? There are so many. One way is to employ a common modeling strategy, called *salami tactics,* in which a complex problem is broken into smaller, more manageable parts (Starfield, Smith, and Bleloch, 1994). The "slices" that I'll use are the phases in a typical project life cycle—planning, organizing, staffing, directing, and controlling.

Project Manager's Role over the Project Life Cycle

Planning

During the planning stage, you as the project manager must establish project objectives and performance requirements. Remember to involve key participants in the process (since, according to an old rule of thumb, involvement builds commitment).

> ### CRITICAL SUCCESS FACTORS AND THEIR IMPORTANCE FOR SYSTEM IMPLEMENTATION
>
> The following factors are listed in decreasing order of correlation.
>
> 1. *Project mission.* Initial clearly defined goals and general directions.
> 2. *Top management support.* Willingness of top management to provide the necessary resources and authority/power for implementation success.
> 3. *Schedule plans.* A detailed specification of the individual action steps for system implementation.
> 4. *Client consultation.* Communication, consultation, and active listening to all parties impacted by the proposed project.
> 5. *Personnel.* Recruitment, selection, and training of the necessary personnel for the implantation project team.
> 6. *Technical tasks.* Availability of the required technology and expertise to accomplish the specific technical action steps to bring the project on-line.
> 7. *Client acceptance.* The act of "selling" final product to its ultimate intended users.
> 8. *Monitoring and feedback.* Timely provision of comprehensive control information at each stage in the implementation process.
> 9. *Communication.* The provision of an appropriate network and necessary data to all key actors in the project implementation process.
> 10. *Troubleshooting.* Ability to handle unexpected crises and deviations from plan.
>
> SOURCE: Pinto, 1986.

Establish well-defined milestones with deadlines. Try to anticipate problems and build in contingencies to allow for them. Carefully outline responsibilities, schedules, and budgets.

ORGANIZING

The first step in organizing is to develop a work breakdown structure that divides the project into units of work. If the project is large and complex, then the next step is to create a project organization chart that shows the structure and relationships of key project members. Finally, schedules, budgets, and responsibilities must be clearly and thoroughly defined.

STAFFING

The major portion of most project successes depends on the people involved with the project. You must define work requirements and, to the extent possible, seek appropriate input when selecting team members. Be sure to orient team members to the big picture of the project. Seek each team member's input to define and agree on scope, budget, and schedule. (Remember, involvement builds commitment, and usually a better product.) Set specific performance expectations with each team member.

Directing

The day-to-day directing of projects involves coordinating project components, investigating potential problems as soon as they arise, and researching and allocating necessary resources. Be sure to remember to display a positive, can-do attitude, and to be available to team members. Recognize team members' good work and guide necessary improvement.

Controlling

Keeping the project on course with respect to schedule, budget, and performance specifications requires paying attention to detail. Some things that usually help are the following:

1. Communicate regularly with team members.
2. Measure project performance by maintaining a record of planned and completed work.
3. Chart planned and completed milestones.
4. Chart monthly project costs.
5. Document agreements, meetings, telephone conversations.

Enormous changes are occurring in the way work and learning are done. You are probably experiencing some of these changes in your classes as you are asked to work on projects in groups and formulate and solve open-ended problems. If you're working at an engineering job, you are surely experiencing some of these changes.

I've tried to provide a perspective on changes that are occurring both in the classroom and in the workplace. One of the most influential references on change is Stephen Covey's *Seven Habits of Highly Effective People* (1989), which has sold millions of copies. Covey lists these habits as follows:

1. *Be pro-active:* Take the initiative and the responsibility to make things happen.
2. *Begin with an end in mind:* Start with a clear destination to understand where you are now, where you're going, and what you value most.
3. *Put first things first:* Manage yourself. Organize and execute around priorities.
4. *Think win/win:* See life as a cooperative, not a competitive arena where success is not achieved at the expense or exclusion of the success of others.
5. *Seek first to understand:* Understand then be understood to build the skills of empathic listening that inspire openness and trust.
6. *Synergize:* Apply the principles of cooperative creativity and value differences.
7. *Renew:* Preserve and enhance your greatest asset, yourself, by renewing the physical, spiritual, mental, and social/emotional dimensions of your nature.

Countless students in my classes have said, "Covey's book changed my life!" The roles of project managers in engineering school and in the workplace are complex and varied. Covey's list provides a good set of heuristics to guide project managers. Another classic that you may want to read is Frederick P. Brooks's *The Mythical Man-Month*. If this book, which is about software project management, intrigues you, then you may want to check out Steve McConnell's *Software Project Survival Guide*. More and more engineers are involved in software development, and these two books will help you manage it.

QUESTIONS

1. What changes have you noted in the workplace or school? Has your school undergone a schedule change recently, for example, from quarters to semesters? How do the changes you've noted compare with those listed in the chapter?

2. This chapter emphasized the changing nature of the workplace and of engineering work, and the needed project manager skills, based on past (and sometimes current) practice. What do you anticipate project manager skills will be in the future? What do the futurists—John Naisbitt, Watts Walker, Esther Dyson, and others—have to say about this?

3. What skills are essential for effective project managers? How can they be enhanced and developed?

4. What does the literature report as keys to project success? How does this list compare with your experience?

5. This chapter organized the project manager's role around the life cycle. Are there other ways that come to mind for organizing the project manager's role? What are they? What are their advantages and disadvantages?

EXERCISE

Interview a project manager, or someone who is involved with project work. Potential interview questions developed by students in my project management classes are listed below. This exercise will not only give you a chance to find out more about project management in practice (and refine your list of essential skills, competencies, and attitudes), but also may help you decide if it is a career path you'd like to pursue.

1. What is the main thrill or interesting reason for being a project manager?

2. Describe a typical work schedule during the week, including number of hours.

3. What directed you to become a project manager? Why not stay in engineering or some other form of business management?

4. Discuss instances of the project manager's role as a team leader and a team member.

5. Discuss the realm and responsibilities of the project manager.

6. What personal characteristics are your best ally during the job's activities?
7. What skills have you had to develop or refine since becoming a project manager?
8. What personal goals are you striving for, personally and professionally?
9. What are your weaknesses that slow you down?
10. What are a few of the frustrations of the job?
11. Describe the project manager's professional credibility and its value.
12. How does the company you work for consider the project manager and his or her responsibilities and opportunities?
13. Describe any battles to complete tasks in the most economical manner and still maintain quality or integrity.
14. How important is the project schedule and what purposes does it serve?
15. Discuss extent of interaction with project owners or clients.
16. What is the project manager's level of involvement with contract negotiations?
17. Discuss the project manager's interaction with other professionals (e.g., engineer or architect).
18. What methods does the project manager use to monitor adequate communication between project owner, architect, client, and the job-site crew?
19. What accounting practices does the project manager use for project budget and regular reviews?
20. Is financial compensation commensurate with the work? What other benefits are there?

REFERENCES

Brooks, Frederick P., Jr. 1995. *The mythical man-month: Essays on software engineering,* anniversary ed. Reading, MA: Addison-Wesley.

Byrne, J. A. 1992. Paradigms for postmodern managers. *Business Week,* Special Issue on Reinventing America, 62–63.

Covey, Stephen R. 1989. *The seven habits of highly effective people.* New York: Simon & Schuster.

Dertouzos, M. L., R. K. Lester, and R. M. Solow. 1989. *Made in America: Regaining the productive edge.* New York: Harper.

Drucker, Peter F. 1993. *Post-capitalist society.* New York: Harper Business.

Gaynor, G. H. 1998. The dawning of a new age: Crossroads of the engineering profession. *Today's Engineer* 1(1): 19–22.

Katzenbach, Jon R., and Douglas K. Smith. 1993. *The wisdom of teams: Creating the high-performance organization.* Cambridge, MA: Harvard Business School Press.

Lientz, Bennet, and Kathryn Rea. 1995. *Project management for the 21st century.* San Diego: Academic Press.

McConnell, Steve. 1998. *Software survival guide: How to ensure your first important project isn't your last.* Redmond, WA: Microsoft Press.

Peters, Thomas J. 1991. *Managing projects takes a special kind of leadership* Seattle Post-Intelligencer, 4/29/1991, p. 64.

Peters, Tom. 1999. The WOW project: In the new economy, all work is project work. *Fast Company, 24,* 116–128.

Pinto, J. K. 1986. Project implementation: A determination of its critical success factors, moderators, and their relative importance across stages in the project life cycle. Unpublished Ph.D. dissertation. Pittsburgh, PA: University of Pittsburgh.

Pinto, J. K., and O. P. Kharbanda. 1995. *Successful project managers: Leading your team to success.* New York: Van Nostrand Reinhold.

Posner, Barry. 1987. Characteristics of Effective *Project Managers. Project Management Journal.* (March)

Smith, Karl A., and Alisha A. Waller. 1997. *New paradigms for engineering education.* Pittsburgh, PA: ASEE/IEEE Frontiers in Education Conference Proceedings.

Starfield, A. M., K. A. Smith, and A. L. Bleloch. 1994. *How to model it: Problem solving for the computer age.* Edina, MN: Burgess.

chapter

6

PROJECT SCHEDULING

Project scheduling is a central yet often overrated aspect of project management. For some the feeling is "We've got a schedule; we're done." Getting a schedule is just one important step in the process of project management. The real work begins when circumstances cause delays and pressures mount to revise the schedule.

INDIVIDUAL REFLECTION   Think about how you typically schedule complex projects, such as completing a major report for a class. Do you make a list of things to do? An outline? Do you draw a concept map? Or do you just start writing?

In this chapter, we'll work our way through the details of the scheduling process. We'll learn the basics of the critical path algorithm and experience firsthand the ideas of forward pass, backward pass, critical path, and float. As you develop an understanding of these concepts and procedures, you will gain insight into managing projects with complex schedules.

Let's revisit the meal-planning exercise from Chapter 4 (p. 51). Take a look at what you did for this exercise. If you didn't do it yet, go back and think about how you would tackle this task. We will use this exercise as the project example throughout this chapter.

REFLECTION   What representation (model) did your group use to determine the time? How did you keep track of which activities had to follow each other and which could be going on at the same time? How did you go about determining the total time the meal preparation and eating would take? Did you make a list? A timeline? Or did you approach the problem in some other way?

Work Breakdown Structure

A common approach for scoping a project is to prepare a work breakdown structure (WBS). The WBS can be presented as a list or an organization chart. A one-level WBS for the meal-planning exercise would be "Prepare the meal," but this wouldn't be too helpful in figuring out what had to be done. A two-level WBS would include:

Preparation
    Boil soup
    Boil rice
    Boil peas
    Brown chicken
    Prepare sauce
    Bake chicken, rice, and sauce
    Open wine and let it breathe
Eating
    Eat soup
    Eat entrée

This two-level WBS provides more specific guidance but still leaves a lot up to the chef (which is OK in many cases).

A more elaborate approach to preparing a WBS is to use Post-it notes to sort out the sequences. There are several possible sequences for the activities for this WBS, depending on how you interpret the "proper order" of preparing this meal. One possible WBS is shown in Figure 6.1.

In this WBS only the activity names and the resource (Pot 1, etc.) that the ac-

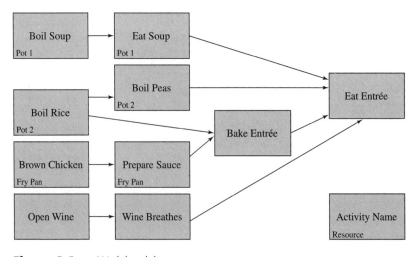

Figure 6.1 Work breakdown structure

tivity uses are listed. Notice that I've made decisions about placing the sauce on the entrée before putting it in the oven, and having the wine with the entrée rather than with the soup. You may have chosen a different sequence, perhaps to have the wine with the soup or to place the sauce on the entrée after it is served. Later we'll explore how these choices affect the schedule.

> **REFLECTION** Have you used the WBS idea for scoping projects? If not, are there places in your personal and professional life where you can immediately apply the WBS idea? How about engineering course or design projects you're working on? If you want more practice, try the office remodeling project exercise at the end of this chapter.

CRITICAL PATH METHOD

Now that we have a WBS for the meal-planning project, we can determine the minimum time to complete it. To do this, we go through Figure 6.1, number each activity and list the time it takes (see Figure 6.2). Examine the precedence network in Figure 6.2 to determine the minimum time to complete the project. Sum the individual activity durations along each path; for example, path 1 + 5 + 10 is 35 + 15 + 25 = 75 minutes. Which path is longest?

Provided that the number of activities is not too large, problems of this type can often be solved by hand. By sketching the relationships between the individual tasks, and taking into account the amount of time each requires for completion, we can determine the total amount of time needed to get the whole process completed.

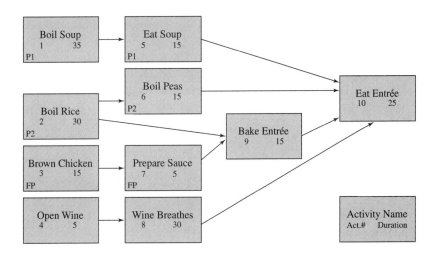

Figure 6.2 Precedence network

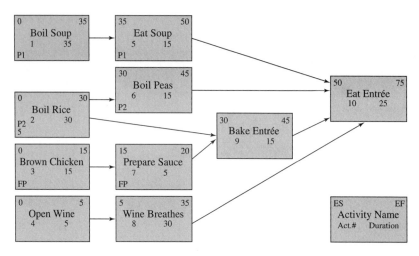

Figure 6.3   Forward Pass

When the number of tasks gets large—say, over 20—then it's quite challenging to keep track of everything by hand. A simple and systematic way of doing this is provided by the critical path method (CPM). This method represents the flow of tasks in the form of a network. To use it, we simply have to know the duration of each of the activities, and the predecessors of each—that is, the set of activities that must have terminated before an activity can begin.

FORWARD PASS—EARLY START (ES) AND EARLY FINISH (EF)

The first step in the CPM is to run through the network from beginning to end and mark the earliest time that each activity can start. In Figure 6.3, this time is in the upper left-hand corner of each activity. This is clearly obtained by adding the earliest start of its latest starting predecessor to that predecessor's duration. When two or more activities must be completed before the next one can start (such as Brown Chicken and Prepare Sauce before Bake Entrée), then the maximum must be used. The early finish (EF) time is determined by summing the early start (ES) and the duration (see Figure 6.3).

BACKWARD PASS—LATE START (LS) AND LATE FINISH (LF)

Similarly, a backward pass is made, establishing the latest possible starting time (late start, LS) that an activity can have, which is the latest start of the earliest starting successor, less the duration of the activity under consideration (see Figure 6.4). The result is the late finish (LF) time.

CRITICAL PATH METHOD

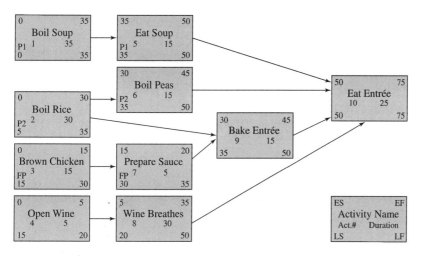

Figure 6.4   Backward Pass

CRITICAL PATH

Activities for which the earliest and latest times turn out to be equal are called *critical*. That is, they cannot be delayed without delaying the duration of the entire project. The path that these activities lie on in the network is known as the *critical path*. The remaining *noncritical activities* have some float (sometimes referred to as slack) and can have their durations increased by some amount before they would become critical and delay the total duration.

FLOATS

The amount by which termination of a noncritical activity can be delayed before it causes one of its successors to be delayed is called the *free float* of that activity. Technically, the free float (FF) is based on early start (ES) times and for any activity i is equal to the minimum early start for activities following activity i minus the early start for i minus the duration (D) for i. Algebraically, the free float is determined as follows:

$$FF_i = (ES_{i+1})_{min} - ES_i - D_i$$

The amount of slack an activity has before it would cause the path on which it lies to become critical is called the *total float*. The total float of an activity is the minimum (out of all of the paths on which it lies) of the sum of its free float and those of all activities ahead of it on the path. Thus an activity is critical if its total float is zero. Technically, the total float (TF) is the difference between the late start (LS) times and the early start (ES) times. Algebraically, the total float for an activity i is determined as follows:

CHAPTER 6 • PROJECT SCHEDULING

$$TF_i = LS_i - ES_i - LF_i - EF_i$$

The numerical solution to the meal-planning problem is given in Table 6.1.

Table 6.1 Meal-planning exercise: critical path method results

Activity	Name	Duration	Resources	Early Start	Early Finish	Late Start	Late Finish	Float Total	Float Free	Current Start	Critical Path
1 [4]	Open Wine	5	1	0	5	15	20	15	0	0	No
2 8	Wine Breathes	30	1	5	35	20	50	15	15	5	No
3 1	Boil Soup	35	1	0	35	0	35	0	0	0	Yes
4 5	Eat Soup	15	1	35	50	35	50	0	0	~35	Yes
5 2	Boil Rice	30	1	0	30	5	35	5	0	0	No
6 6	Boil Peas	15	1	30	45	35	50	5	5	30	No
7 3	Brown Chicken	15	1	0	15	15	30	15	0	0	No
8 7	Prepare Sauce	5	1	15	20	30	35	35	15	~15	No
9 9	Bake Entrée	15	1	30	45	35	50	5	5	30	No
10 10	Eat Entrée	25	1	50	75	50	75	0	0	50	Yes

GANTT CHARTS AND CRITPATH

Another common model for representing scheduling projects is a time-scaled network, called a Gantt chart, where the activities have been laid out on a time axis. The table and Gantt chart for the first eight activities, shown in Figure 6.5, were prepared using the CritPath software program, which is available for downloading from the

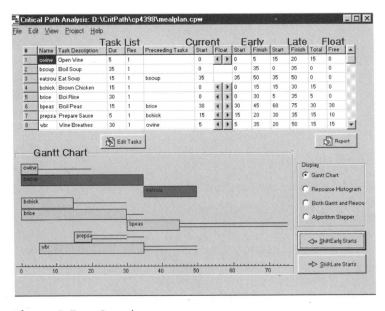

Figure 6.5 Gantt chart

CRITICAL PATH METHOD

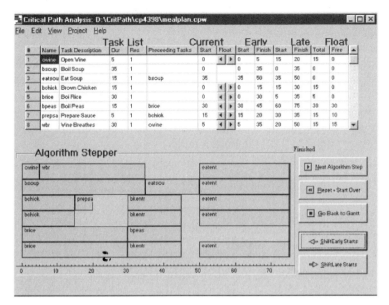

Figure 6.6 Algorithm stepper

BEST Web site and is bundled with *How to Model It* (Starfield, Smith, and Blelock, 1994). The CritPath program is set up to display only eight activities at a time. If you want to view the rest of it, download the CritPath program and play with it.

If you are still having difficulty understanding the differences between free and total float or are struggling with the critical path calculations, use the algorithm stepper in the CritPath program. It will walk you through the process using a graphical representation (see Figure 6.6).

Notice how there is no gap in the path that includes the activities Open Wine, Eat Soup, and Eat Entrée? That means, of course, that they are on the critical path. Also notice how there is a gap after the activities Wine Breathes, Prepare Sauce, Bake Entrée, and Boil Peas; this means they have free float in addition to having total float. The activities Brown Chicken and Open Wine are followed by a gap farther down the path, but not by an immediate gap; this means that they have total float but not free float.

INDIVIDUAL REFLECTION Take a few minutes to think about the advantages and disadvantages of the two representations of the meal-planning project—the precedence network (Figure 6.4) and the Gantt chart (Figure 6.5). What are the unique features of each? What specific features does each represent? Where is each appropriate?

In the above Reflection, you may have concluded that both the precedence network and the Gantt chart are essential for understanding complex projects and communicating project information. The Gantt chart is a time-scaled network, since time is represented directly. It gives a clear picture of the duration of events, but it doesn't directly show the nature of the interdependence, that is, what has to follow what. The precedence network, on the other hand, clearly shows the interdependence—precedence and simultaneity, that is, what has to follow what and what can be going on at the same time—but it doesn't directly show the time required for each activity.

The CritPath program, like most project scheduling software, uses the precedence network representation to do the critical path calculations. Many people find the Gantt chart most useful for tracking project progress.

CRITICAL PATH METHOD SUMMARY

In summary, the sequence of steps to apply the critical path method to project scheduling is as follows:

1. Develop a work breakdown structure (WBS).
2. Connect the activities in the WBS by arrows that indicate the precedence.
3. Perform the critical path analysis calculations either by hand (for a simple problem) or with computer software (for a complex problem).
4. Create graphical representations—a precedence network and a Gantt chart—that suit your purposes.

BUS SHELTER CONSTRUCTION EXAMPLE

Now that we've worked through the meal-planning exercise in some detail, let's try another example. Consider the construction project outlined in Table 6.2. Note that the precedence relationships are specified so you don't have to create a work breakdown structure; however, developing a precedence network is an important step. The resources column specifies the number of people required for each task.

Table 6.2 Bus shelter construction example

Job	Name	Duration	Resources	Predecessors
1	Shelter Slab	2	2	5
2	Shelter Walls	1	1	1
3	Shelter Roof	2	2	2, 4
4	Roof Beam	3	2	2
5	Excavation	2	3	
6	Curb and Gutter	2	3	5
7	Shelter Seat	1	2	4, 6
8	Paint	1	1	7
9	Signwork	1	2	2, 6

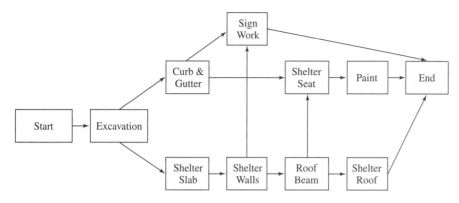

Figure 6.7 Precedence network for bus shelter

Determine the minimum time required to complete the bus shelter. Develop a precedence network. Identify the critical path. Draw a Gantt chart for the project. Give it a try before looking further. (A precedence network is sketched out in Figure 6.7.)

Next, perform the critical path analysis calculations. You may do this by hand if you want more practice, or you can use CritPath or another commercial project-scheduling program. The table and Gantt chart for the first eight activities are shown in Figure 6.8.

#	Name	Task Description	Dur	Res	Preceeding Tasks	Start	Float	Start	Finish	Start	Finish	Total	Free
1	shslab	Shelter Slab	2	2	excv	2		2	4	2	4	0	0
2	shwall	Shelter Walls	1	1	shslab	4		4	5	4	5	0	0
3	shroof	Shelter Roof	2	2	shwall rfbm	8		8	10	8	10	0	0
4	rfbm	Roof Beam	3	2	shwall	5		5	8	5	8	0	0
5	excv	Excavation	2	3		0		0	2	0	2	0	0
6	cbgut	Curb & Gutter	2	3	excv	2		2	4	6	8	4	1
7	shseat	Shelter Seat	1	2	rfbm cbgut	8		8	9	8	9	0	0
8	paint	Paint	1	1	shseat	9		9	10	9	10	0	0

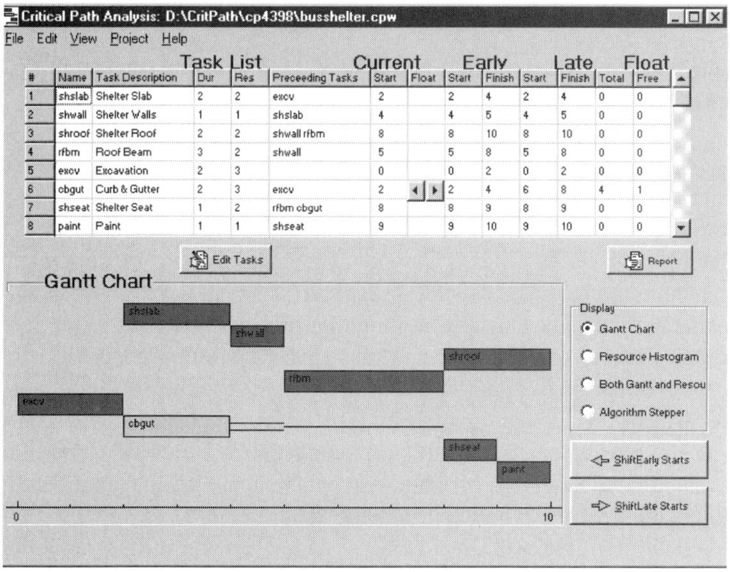

Figure 6.8 Gantt chart for bus shelter

Table 6.3 Bus shelter construction: critical path method results

Activity	Name	Duration	Resources	Early Start	Early Finish	Late Start	Late Finish	Float Total	Float Free	Current Start	Critical Path
1	Shelter Slab	2	2	2	4	2	4	0	0	2	Yes
2	Shelter Walls	1	1	4	5	4	5	0	0	4	Yes
3	Shelter Roof	2	2	8	10	8	10	0	0	8	Yes
4	Roof Beam	3	2	5	8	5	8	0	0	5	Yes
5	Excavation	2	3	0	2	0	2	0	0	0	Yes
6	Curb and Gutter	2	3	2	4	6	8	4	1	2	No
7	Shelter Seat	1	2	8	9	8	9	0	0	8	Yes
8	Paint	1	1	9	10	9	10	0	0	9	Yes
9	Signwork	1	2	5	6	9	10	4	4	5	No

Look carefully at the critical path method results and the Gantt chart. Notice that there is more than one critical path. The presence of multiple critical paths can be seen on the Algorithm Stepper feature of the CritPath program, since all the paths through the network are displayed. Get the CritPath software and try it. Table 6.3 shows the entire set of results for the bus shelter construction project.

Try developing and setting up your next project using the critical path method. Do the calculations by hand a couple times to familiarize yourself with the forward pass and backward pass of the algorithm. Then use CritPath or another commercial project-scheduling program.

PROJECT RESOURCE AND COST CONSIDERATIONS

RESOURCE LEVELING

Critical activities, having no slack, cannot be extended or shifted without upsetting the scheduled completion of the project. However, the slack afforded by noncritical activities can be exploited to provide the best distribution of resources over the duration of the entire project. For example, it might be difficult or expensive to hire more than a certain number of laborers at any one time. By shifting noncritical activities within their floats, it is possible to spread the distribution of labor more evenly over the span of the project. At other times it may be beneficial to load the distribution in a certain way, for example, if work over a holiday period is to be minimized. Decisions of these kinds can be made only when the constraints (e.g., earliest and latest start times) of the schedule have been determined. Look at what happens to the resource histogram for the bus shelter project when you shift all activities to their early start times (Figure 6.9). What do you expect would happen if you shifted all activities to their late start times? See Figure 6.10.

Project Resource and Cost Considerations

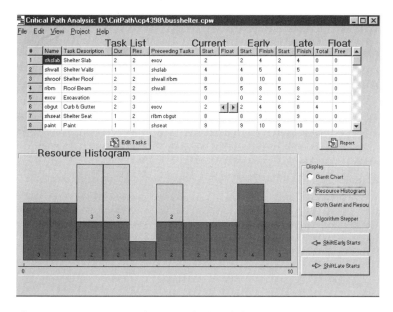

Figure 6.9 Resource histogram for bus shelter

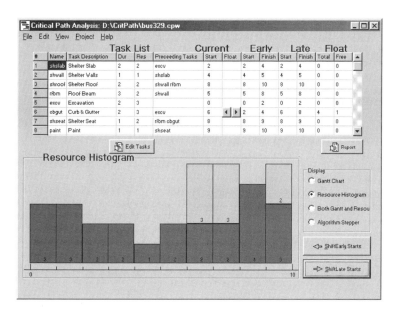

Figure 6.10 Resource histogram—latest times

Compare Figures 6.9 and 6.10 to help you decide what the best allocation of resources would be. Notice that there is not too much that you can do to level the resources in this case, which is sometimes the situation with real projects.

COST CONSIDERATIONS

You may on some projects work through the critical path calculations and find that the required project duration is greater than the time you have available. Or, more commonly, you may get behind during the project due to bad weather, late deliveries, work delays, and so on. In such cases, you could, of course, go to your supervisor or professor and ask for an extension. Alternatively, you could add resources (e.g., people, overtime) to activities on the critical path to decrease their duration, thus decreasing the time for the entire project. Technically, this is known as "crashing" a project. Why wouldn't we add resources to noncritical activities? Let's work through the example shown in Table 6.4 to get a better sense of how this works.

Table 6.4 Critical path method cost example

Task	Precedence	Normal Time	Crash Time	Normal Cost	Incremental Cost Per Day
A	—	8	4	$ 800	$300
B	—	4	3	600	100
C	B	2	1	1,000	400
D	A, C	3	2	300	200

Figure 6.11 shows the normal schedule duration. The normal schedule cost is just the sum of the normal costs for the four activities—$2,700.

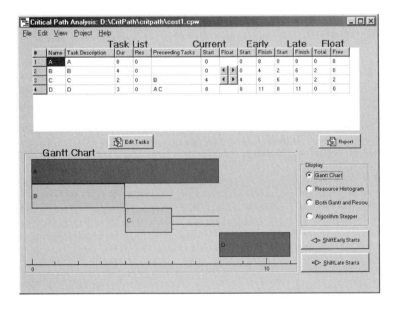

Figure 6.11 Gantt chart for cost example

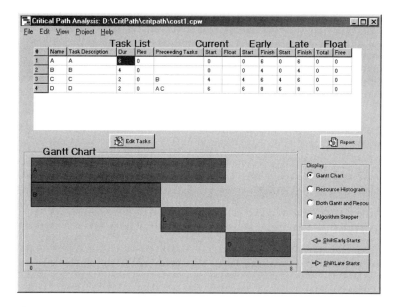

Figure 6.12 Gantt chart—all paths critical

As the project manager you would have to choose which activities along the critical path to add resources to, in order to decrease their duration. What criteria would you use to choose the activities? Ordinarily, the project manager would crash the activities that had the minimum cost per unit of time saved. Often convenience, availability, and other factors must be considered.

Let's look at what happens as we decrease the duration of activities on the critical path for the example above. The lowest cost choice is Activity D. By decreasing the duration of Activity D from 3 to 2, the overall project duration decreases to 10 and the cost increases to $2,700 + $200 = $2,900. Next we can add resources to Activity A to decrease it to 7. The overall project duration becomes 9, and the cost increases to $3,000. Adding more resources to A to decrease it to 6 decreases the overall project duration to 8 and increases the cost to $3,300. The updated Gantt chart is shown in Figure 6.12. We can continue to decrease the duration of the overall project, but now we must add resources to more than one activity and hence the cost increases at a higher rate.

As you add resources to critical activities and decrease the duration along the critical path, eventually, more and more activities become critical.

INDIVIDUAL AND GROUP REFLECTION What are some good strategies for using the float over the life of a project? Do you, for example, let things slide early on, thus using the float up early? Or do you wait until later in the project to use the float? As a project manager, how do you recommend that the float be utilized?

The Role of Computer-based Project Management Software

It is not hard to see that if we were to add a few more activities, the problem would soon become unmanageable by hand. Further, if changes have to be made either to the order in which activities must occur, or in the time in which they can be completed, the entire process would have to be repeated. The advantage of the critical path method is that it is indeed systematic. It can be described as a formal set of instructions that can be followed by a computer. Alterations in the data can be made repeatedly and the problem quickly solved, again by the machine. This would enable us to obtain the benefit of what-if analysis, the process of making changes and seeing the effects of those changes immediately. Such analysis gives the user an intuitive feel for the problem. The role of the computer will be addressed in more detail in Chapter 9.

Questions

1. What is a work breakdown structure (WBS)? Why is it important? What are some of the types of WBSs?
2. What is the critical path method (CPM)? What are free and total float?
3. Explain the differences between resource leveling and crashing a project.
4. How have you typically managed complex projects in the past? How well did the approach work? Not all projects merit taking the time to develop a schedule, since they can be managed by making a list or by using your day planner. Where could you apply the critical path method and how do you expect it might help?
5. How does the CritPath program compare with other software packages you've used? With a spreadsheet, for example?
6. What are some of the advantages and disadvantages of relying on project-scheduling algorithms?

Exercises

1. Work Breakdown Structure: Office Remodeling Project

The following activities must be accomplished to complete an office remodeling project:

Activity	Estimated Duration (Days)
Procure paint	2
Procure new carpet	5
Procure new furniture	7
Remove old furniture	1
Remove old carpet	1
Scrub walls	1
Paint walls	2
Install new carpet	1
Move in new furniture	1

1. Create a possible work breakdown structure (WBS) for the remodeling project.
2. When can the new furniture be moved in?
3. What is the minimum project duration?
4. Which activities do you have to pay close attention to if you want to finish at the earliest possible time?

2. Resource Leveling

Given the following set of project data, determine the smoothest distribution of resources. Assume resources are transferable.

Task	Duration	Resources	Predecessor
1	7	8	
2	5	6	
3	4	4	2
4	2	4	1, 3
5	3	6	2
6	1	6	5

3. Crashing

Given the following project data, determine the normal schedule cost. Crash the project as far as possible. List the project duration and cost for each step along the way.

Task	Precedence	Normal Time	Crash Time	Normal Cost	Incremental Cost Per Day
A	—	5	3	$ 500	$300
B	A	4	2	600	100
C	A	6	4	1,000	400
D	B, C	3	2	300	200

4. Scheduling Case Study

Develop a work breakdown structure, precedence network, and Gantt chart for a project you're involved with. Complete the critical path analysis calculations. Use these representations to guide the project and to review progress.

5. A Minnesota Dream Project—Lakeside Cabin Cost Estimate

Imagine you've just inherited a lakeside lot in Northern Minnesota (beautiful in the summer) that has an access road, electricity, and water and sewage disposal. You've also been given a modest sum with which to buy building materials. If you're having trouble imagining yourself being so lucky, consider the possibility that a friend or neighbor asked you for an estimate on how much it would cost to build a cabin and for help in designing and building. Engineers are often expected to know such things. How much would it cost for a modest, say 24-foot-square cabin? How long would it take you to build it?

The cabin cost estimate project is a favorite of students in my project management classes. Since we haven't done too much with cost estimating (it's usually done in engineering economics courses), I'll provide a little guidance. I suggest you take the following steps:

1. Guess. Take a wild guess at what it would cost for materials for a modest cabin.
2. Look up the square-foot costs for typical residential construction ($50–$70 per square foot) and adjust it down for a modest cabin. Many students use $30 per square foot for a modest cabin.
3. Use a unit cost approach:
   a. Develop a floor plan.
   b. Create a detailed list of materials with associated quantities.
   c. Find the cost of the materials in a local building center brochure (or cost manual).
   d. Use a spreadsheet to list and calculate the total cost of the materials.
4. Compare the three cost estimates—guess, cost per square foot, and unit cost.

Use the procedures outlined in this chapter to determine how long it would take you to build it; that is, develop a WBS and a schedule using the critical path method.

Reference

Starfield, A. M., K. A. Smith, and A. L. Bleloch. 1994. *How to model it: Problem solving for the computer age.* Edina, MN: Burgess. (Includes disk with CritPath and WinExp.)

chapter

7

PROJECT MONITORING AND EVALUATION

Two important parts of project management are keeping on top of projects through monitoring and conducting end-of-the-project reviews through evaluation. There are many ways to keep informed, keep others informed, and coordinate projects. Informal one-on-one meetings, e-mail, and phone conversations are the most common. Other simple forms include fax messages, voice messages, and handwritten notes. More involved forms of information exchange are groupware (such as Lotus Notes), informal reports, formal memos and meetings, and formal presentations. Each has advantages and disadvantages.

REFLECTION What is your experience with face-to-face, phone, fax, e-mail, and handwritten communications? Under what conditions do you prefer each?

Tom Peters (1989) popularized the idea of management by wandering around (MBWA), which has become an important aspect of most project managers' daily lives. Effective project managers stay in touch. More recently, Peters (1999) has said that "all work is project work" and that individuals are identified and recognized by their project work.

This chapter begins by discussing meetings and progresses through successively more complex ways to monitor the work of projects and groups before finishing with strategies for evaluating projects.

MEETINGS

Although meetings are a principal way of keeping up-to-date, they are also one of the alleged banes of project managers and many others. Dressler (1996) lists some of the major complaints people have about meetings:

- Their purpose is unclear.
- Participants are unprepared.
- Key people are absent or missing.
- The conversation veers off track.
- Participants don't discuss issues but instead dominate, argue, or take no part at all.
- Decisions made during the meeting are not followed up.

INDIVIDUAL REFLECTION Think about some of your best (and worst) experiences in meetings. How do the complaints listed above compare with your experiences in meetings? What additional challenges have you encountered? What conditions contribute to best and worst experiences? What are some of the strategies you use or have seen used to counter the worst experiences?

Five guidelines from the book *Meetings: Do's, Don'ts, and Donuts* (Lippincott, 1994) can help alleviate some of the most common problems:

1. State in a couple of sentences exactly what you want your meeting to accomplish.
2. If you think a meeting is the best way to accomplish this, then distribute an agenda to participants at least two days in advance.
3. Set ground rules to maintain focus, respect, and order during the meeting.
4. Take responsibility for the meeting's outcome.
5. If your meeting isn't working, try other tools, such as brainstorming.

Using a meeting process such as the one outlined in *The Team Handbook* (Scholtes, Joiner, and Streibel, 1996) can help. The authors describe the three-part meeting process as follows:

Before

Plan.

Clarify meeting purpose and outcome.

Identify meeting participants.

Select methods to meet purpose.

Develop and distribute agenda.

Set up room.

During

Start: check-in, review agenda, set or review ground rules, clarify roles.

Conduct—cover one item at a time, manage discussions, maintain focus and pace.

How to Run a Meeting

Plan the Meeting
- Be clear on objectives of the meeting.
- Be clear why you need the meeting.
- List the topics to be addressed.

Inform
- Make sure everyone knows exactly what is being discussed, why and what you want from the discussion.
- Anticipate what people and information may be needed and make sure they are there.

Prepare
- Prepare the logical sequence of items.
- Allocate time on the basis of importance, not its urgency.

Structure and Control
- Put all evidence before interpretation and action.
- Stop people from jumping back and going over old ground.

Summarize and Recall
- Summarize all decisions and record them with the name of the person responsible for any action.

SOURCE: Adapted from the video "Meetings, Bloody Meetings," starring John Cleese 1993.

Close—summarize decisions, review action items, solicit agenda items for next meeting, review time and place for next meeting, evaluate the meeting, thank participants.

After

Distribute or post meeting notes promptly.

File agendas, notes, and other documents.

Do assignments.

The box "How to Run a Meeting" also provides good advice for running effective meetings.

Monitoring Group Effectiveness

One common method for monitoring the effectiveness of group work is the plus/delta group processing approach typically attributed to the Boeing Company. Near the end of the meeting the group stops working on the task and spends a few minutes discussing how well they worked. The group makes a list that records what went well on one side (+) and what they could do even better on the other (Δ). Other methods include individual reflection, using instruments such as the one in Figure 7.1. Members fill in the chart and then discuss each other's scores and comments.

More complex monitoring involves collecting data on individual participation in the team. Several observation forms are available. The one I start students with in my project management classes is shown in Figure 7.2. Any task or maintenance behavior (see Chapter 3) may be listed in the rows. Group members take turns observ-

For each trait, rate the team on a scale of 1 to 5:

1 = Not present (opposite trait present)
2 = Very poor (not much evidence of positive trait)
3 = Poor (some positive trait seen)
4 = Good (positive trait evident more than opposite trait)
5 = Very good (large amounts of evidence of positive trait)

Positive Trait	Score	Comments
Safety		
Inclusion		
Free interaction		
Appropriate level of interdependence		
Cohesiveness		
Trust		
Conflict resolution		
Influence		
Accomplishment		
Growth		

SOURCE: Uhlfedler, 1997.

Figure 7.1   A sample individual reflection instrument

ing the group and recording each member's participation. They then provide feedback about the group's functioning during the processing phase. A rule of thumb that I commonly use is "Keep the feedback specific, descriptive, immediate, and positive." I give negative feedback only if the person requests it. If negative feedback is requested, the person is usually ready to hear it. Only then will it be helpful. More rules include the following, from Scholtes (1988):

Guidelines for Constructive Feedback

- Acknowledge the need for feedback.
- Give positive feedback. Give negative feedback only if the recipient asks for it.
- Understand the context.
- Know when to give feedback.
- Know how to give feedback:
  Be descriptive.
  Don't use labels.
  Don't exaggerate.
  Don't be judgmental.
  Speak for yourself.
  Talk first about yourself, not about the other person.
  Phrase the issue as a statement, not a question.
  Restrict your feedback to things you observed.
  Help people hear and accept your compliments when giving positive feedback.

Observation Category	Names				Total
Task *Contributes ideas*					
Maintenance *Encourages*					
Total					

Notes:

Observation Directions:

1. Move your chair so you can see each member of the group clearly without interacting with them.
2. Write the name of each person in the group at the top of one of the columns on the observation sheet.
3. Watch each person systematically. As you see each person display one of the two behaviors specified (contributes ideas or encourages), place a hatch mark below his or her name, in the box to the right of the appropriate behavior.
4. *Task* means anything that helps the group accomplish its task. For example, *Contributes ideas* means giving an idea related to one of the questions on the worksheet and/or to something said by another group member related to the task.
5. *Maintenance* means anything that helps improve working relationships in the group. *Encourages* means praising others' ideas or inviting others to contribute.
6. You may make some notations below the grid that may help you explain some scoring.
7. When time is called, total the hatch marks in each column and across each row.
8. When you give feedback to the group you will only give them the column totals and the row totals. Let them see the sheet. Do not interpret what you saw, or what the totals might suggest. You may be tempted to soften what you say. Don't allow this to happen. You are not criticizing; you are only reporting what you saw, related to very specific behaviors.
9. It is the job of the group to discuss what the totals might suggest about how they functioned as a group, and to develop one or two sentences that capture this thought.

Figure 7.2   Observation Sheet

Behavior						Totals	%
Initiating							
Proposing							
Building							
Reacting							
Supporting							
Disagreeing							
Defending/attacking							
Clarifying							
Testing understanding							
Summarizing							
Seeking information							
Giving information							
Totals							
Process							
Shutting out							
Bringing in							

Figure 7.3   Xerox's Interactive Skills Coding Worksheet

Another form that I've found very useful is shown in Figure 7.3. This was developed by Xerox Corporation and is included in the company's interactive skills workbook (1986). On facing page is a table that defines and gives examples of behavior categories listed in the worksheet.

The purpose of collecting data on group functioning by observing and other means is not only to provide data for monitoring but also to help each member attend to how the group is performing. Even if you don't have an opportunity to systematically observe a team, read the definitions and examples in Figure 7.3 and think about how you can expand your repertoire of behaviors.

TEAM TALK ANALYSIS

An even more sophisticated approach to processing the work of teams was developed by Donnellon (1996), who claims, "Not all groups are teams." Although, as I mentioned in Chapter 2, the words *group* and *team* are often used interchangeably, it is important to distinguish between the gathering of people into groups and the purposeful formation of a team. A *team,* according to Donnellon, is "a group of people who are necessary to accomplish a task that requires the continuous integration of the expertise distributed among them" (p. 10).

Donnellon studied team talk and devised six dimensions along which to assess teams: identification (with what group team members identify); interdependence (whether team members felt independent from or interdependent with one another);

Definitions and Examples

Behavior Category	Definition	Examples
Proposing	A behavior that puts forward a new suggestion, proposal, or course of action.	"Let's deal with that one tomorrow." "I suggest we add more resources to . . ."
Building	A behavior that usually takes the form of a proposal, but that actually extends or develops a proposal made by another person.	". . . and your plan would be even better if we added a second reporting stage" "If I can take that further, we could also use the system to give us better cost control."
Supporting	A behavior that makes a conscious and direct declaration of agreement with or support for another person, or for their concepts and opinions.	"Yes, I go along with that." "Sounds OK to me." "Fine. I accept that."
Disagreeing	A behavior that states a direct disagreement or which raises obstacles and objections to another person's concepts or opinions. Disagreeing is about issues.	"No, I don't agree with that." "Your third point just isn't true." "What you're suggesting won't work."
Defending/attacking	A behavior that attacks another person, either directly or by defensiveness. Defending/attacking usually involve value judgments and often contain emotional overtones. They are usually about *people,* not issues.	"That's stupid." ". . . and your third point is either stupid or an out-and-out lie." "Don't blame me, it's not *my* fault; it's *John's* responsibility."
Testing understanding	A behavior that seeks to establish whether or not an earlier contribution has been understood. It differs from seeing information in that it is an attempt to ensure agreement or consensus of some kind, and refers to a prior question or issue.	"Can I just check to be sure we're talking about the same thing here?" "Can I take it that we all now agree on this?"
Summarizing	A behavior that summarizes or restates, in a compact form, the content of previous discussions or events.	"So far, we have agreed (*a*) To divide responsibilities, (*b*) To meet weekly (*c*) To finish the draft proposal by . . ."
Seeking information	A behavior that seeks facts, opinions, or clarification from another person.	"What timeline did we agree to?" "Can anyone tell me what page that table is on?" "Have you checked that thoroughly?"
Giving information	A behavior that offers facts, opinions, or clarification to other people.	"I remember a case like that last year." "There are at least three alternatives."
Bringing in*	A behavior that invites views or opinions from a member of the group who is not actively participating in the discussion.	"Dick, have you anything to say on this one?" "Cheryl has been very quiet. I wonder whether she has anything she would like to say here."
Shutting out*	A behavior that excludes another person or persons, or reduces their opportunity to contribute. Interrupting is the most common form of shutting out.	Jose: "What do you think, Bob?" Karl: "What I think is . . ." (Karl is shutting out Bob.)

Figure 7.3 Xerox's Interactive Skills Coding Worksheet (Continued)

*Characteristic of a process behavior.

power differentiation (how much team members use the differences in their organizational power); social distance (whether team members feel close to or distant from one another socially); conflict management tactics (whether members use the tactics of force or collaboration to manage their conflicts); and negotiation process (whether the team uses a win–lose or a win–win process). Donnellon then used these dimensions to differentiate between nominal teams and real teams, as shown in Table 7.1.

Table 7.1 Nominal versus real groups

	Nominal Team	Real Team
Identification	Functional group	Team
Interdependence	Independence	Interdependence
Power differentiation	High	Low
Social distance	Distant	Close
Conflict management tactics	Forcing, accommodating, avoiding	Confronting, collaborating
Negotiation process	Win–lose	Win–win

SOURCE: Donnellon, 1996.

The dimensions shown in the table are consistent with the underlying conceptual framework in this book. I encourage you to examine the groups and teams you're in along these dimensions. Donnellon's team talk audit for assessing team dynamics is included in her book. Use this instrument to attend to the team talk, reflect on what it tells you about the team, and then plan how you will discuss the assessment with the team.

Donnellon also described five types of teams based on the categorization in her six dimensions. In Table 7.2, I've summarized some of the more direct paths. Think about where your team fits.

Table 7.2 Team talk dimensions

Identification	Interdependence	Power Differentiation	Social Distance	Negotiation	Conflict Management	Profile
Team	High	Low	Close	Win–win	Collaborative	Collaborative
Team	Moderate	Low	Close	Win–win	Force–avoid	Mostly collaborative
Team	Moderate	High	Close			Emergent
Both	High	High	Distant	Win–lose	Force–avoid	Adversarial
Both	Low	High	Distant	Win–lose	Force–avoid	Adversarial
Function	Low	Low	Distant			Nominal
Function	Low—independence	High	Distant	Win–lose	Force–avoid	Doomed

SOURCE: Donnellon, 1996.

Donnellon's work indicates that there are very few paths to collaborative team profiles, a conclusion borne out in the work of Katzenbach and Smith (1993), and Bennis and Beiderman (1997), whose case studies note that very few teams perform at the highest levels. With these dimensions in mind, carefully examine your group and team experiences and then explicitly discuss the performance (functioning) of your team to help you decide (1) to leave if your team is doomed, (2) refine the team if you're in the middle, or (3) celebrate and continue performing if you're a collaborative team.

Using the monitoring and processing formats described above to systematically reflect on the team's performance on both task and maintenance dimensions will help the team achieve its goals and help the members get better at working with one another. Group processing takes time and commitment, and is typically difficult for highly motivated, task-oriented individuals. Spending a little bit of carefully structured time on how the team is functioning can make an enormous difference in the team's effectiveness and quality of the working environment.

Project Evaluation

At the end of a project it is important, and often a requirement, to conduct an evaluation. Typically, a set of project evaluation questions guides this process. The following, generated by Haynes (1989), is a typical set of questions:

1. How close to scheduled completion was the project actually completed?
2. What did we learn about scheduling that will help us on our next project?
3. How close to budget was final project cost?
4. What did we learn about budgeting that will help us on our next project?
5. Upon completion, did the project output meet client specifications without additional work?
6. What, if any, additional work was required?
7. What did we learn about writing specifications that will help us on our next project?
8. What did we learn about staffing that will help us on our next project?
9. What did we learn about monitoring performance that will help us on our next project?
10. What did we learn about taking corrective action that will help us on our next project?
11. What technological advances were made?
12. What tools and techniques were developed that will be useful on our next project?
13. What recommendations do we have for future research and development?
14. What lessons did we learn from our dealings with service organizations and outside vendors?
15. If we had the opportunity to do the project over, what would we do differently?

Continual Evaluation

Evaluation doesn't have to occur only at the end of the project; it is often initiated when a project falls behind schedule or goes over budget. You have probably been involved in group projects that got behind schedule or used more resources that were initially allocated.

> **INDIVIDUAL REFLECTION**   How have you dealt with projects that get behind schedule or use more resources than were initially allocated? What are some of the strategies you've used? Take a few minutes and reflect on dealing with delays and cost overruns.

There are lots of internal things you can do with your project team to address delays and resource excesses. Sometimes it's necessary to try to change external conditions to address delays and overruns. Here's a list of some things you can do:

1. *Recover later in the project.* If there are early delays or overruns, review the schedule and budget for recovery later. This is a common strategy in many projects. How often have you done extraordinary work at the last minute, especially the night before the project is due?
2. *Reduce project scope.* Consider eliminating nonessential elements or containing scope creep. Engineers often find better ways of doing things during projects and are sometimes perfectionists, so there is a tendency for the scope to creep.
3. *Renegotiate.* Discuss with the client the possibility of extending the deadline or increasing the budget. How often have you asked a professor for an extension? This is a common strategy, but sometimes there is not flexibility.
4. *Add resources.* Adding resources—people, computers, and so on—to a project (activities along the critical path, as you learned in Chapter 6) can reduce the duration. The increased costs must be traded off with the benefits of the reduced schedule.
5. *Offer incentives or demand compliance.* Sometimes by offering incentives (provided you don't endanger peoples' lives or sacrifice performance specification) you can get a project back on track. Other times you may have to demand that people do what they said they would do.
6. *Be creative.* If the delay is caused by resources that have not arrived, you may have to accept substitutions, accept partial delivery, or seek alternative sources.

Building Quality into Projects

Evaluation and continual improvement often become an ongoing part of projects and company culture. Often this aspect of company or organizational culture is described

Table 7.3 Thinking about quality

Old	New
Competition motivates people to do better work	Cooperation helps people do more effective work
For every winner there's a loser	Everyone can win
Please your boss	Please your customer
Scapegoating pinpoints problems	Improve the system
Focus improvements on individual processes	Focus on the purpose of the overall system, and how the processes can be improved to serve it better
Find the cause and fix the problem	First, acknowledge there is variation in all things and people; see if the problem falls in or outside the system
The job is complete if specifications have been met	Continual improvement is an unending journey
Inspection and measurement ensure quality	A capable process, shared vision and aim, good leadership, and training are major factors in creating quality
Risks and mistakes are bad	Risks are necessary and some mistakes inevitable when you practice continual improvement
You can complete your education	Everyone is a lifelong learner
Bosses command and control	Bosses help workers learn and make improvements
Bosses have to know everything	The team with a good leader knows and can do more
Short-term profits are best	Significant achievement in a complex world takes time
You don't have to be aware of your basic beliefs	You must be conscious of your beliefs and constantly examine and test them to see if they continue to be true
Do it now	Think first, then act

SOURCE: Dobyns and Crawford-Mason, 1994.

as a quality initiative. Table 7.3 provides a set of insightful contrasts between old and new thinking about quality.

Engineers are often required to help develop a quality initiative in their organization. You may have been involved in a quality initiative in your work or school. Although there has been some attention paid to quality in schools (see Langford and Cleary, 1995, for example), much of the emphasis on quality has been in the workplace. Business and industry have taken the lead, as indicated, for example, by Ford Motor Company's motto "Quality is job one."

Some quality basics include a systems perspective, emphasis on the customer, and understanding variation. Knowledge of sources of variation, especially ways of measuring and documenting them, and strategies for reducing variability and maintaining consistent quality are essential.

Further reading on quality is contained in the references by Deming (1993, 1996), Bowles and Hammond (1991), Dobyns and Crawford-Mason (1994), Brassand (1989), Sashkin & Kiser (1994), and Walton (1986). You may also want to consult a basic textbook on quality, such as Summers (1997).

QUESTIONS

1. Where did you develop skills for monitoring the work on project teams? Have you had an opportunity to observe a project team? If so, where? What did you learn from the experience? Do you try to attend to what's happening within the group while it is working?

2. What can you do to improve the functioning of teams during "boring and useless" meetings? List things you can do and strategies for doing them. Try them out!

3. Check out some of the ethnographic research on work in organizations, such as Brown and Duguid (1991) article in *Organizational Science*. (The paper is also available on the Xerox Palo Alto Research Center Web site). How does research affect your view of work in organizations?

4. What are some strategies for building quality into projects?

5. How can project evaluation, which is often seen as a punitive process, become a more positive and constructive process? What are things that project team members and managers can do to make evaluation an ongoing part of project work?

REFERENCES

Bennis, W., and P. W. Biederman. 1997. *Organizing genius: The secrets of creative collaboration.* Reading: Addison-Wesley.

Bowles, J., and J. Hammond. 1991. *Beyond quality: How 50 winning companies use continuous improvement.* New York: Putnam.

Brassand, Michael. 1989. *The memory jogger plus: Featuring the seven management and planning tools.* Methuen, MA: GOAL/QPC.

Brown, John Sealy, and Paul Duguid. 1991. Organizational knowledge and communities of practice: Toward a unified view of working, learning, and innovation. *Organizational Science* 2(1): 40–57.

Cleese, J., and A. Jay. 1993. *Meetings, bloody meetings.* London: Video arts.

Deming, W. E. 1986. *Out of crisis.* Cambridge, MA: MIT Center for Advanced Engineering Study.

———. 1993. *The new economics for industry, government, education.* Cambridge, MA: MIT Center for Advanced Engineering Study.

Dobyns, Lloyd, and Clare Crawford-Mason. 1994. *Thinking about quality: Progress, wisdom, and the Deming philosophy.* New York: Times.

Donnellon, Anne. 1996. *Team talk: The power of language in team dynamics.* Cambridge, MA: Harvard Business School Press.

References

Dressler, C. 1996. Cited in Lewis, J. P. 1998. *Team-based project management.* New York: AMACOM.

Haynes, M. E. 1989. *Project management: From idea to implementation.* Los Altos, CA: Crisp Publications.

Katzenbach, Jon R., and Douglas K. Smith. 1993. *The wisdom of teams: Creating the high-performance organization.* Cambridge, MA: Harvard Business School Press.

Langford, David P., and Barbara A. Cleary. 1995. *Orchestrating learning with quality.* Milwaukee: ASQC Quality Press.

Lippincott, S. 1994. *Meetings: Do's, don'ts, and donuts: The complete handbook to successful meetings.* Pittsburgh: Lighthouse Point Press.

Peters, Thomas J. (with Nancy Austin). 1989. *A passion for excellence.* New York: Knopf.

———. 1999. The WOW project: In the new economy, all work is project work. *Fast Company* 24 (May): 116–28.

Sashkin, Marshall, and Kenneth J. Kiser. 1993. *Putting total quality management to work.* San Francisco: Berrett-Koehler.

Scholtes, Peter. 1988. *The team handbook: How to use teams to improve quality.* Madison, WI: Joiner Associates.

Scholtes, Peter R., B. L. Joiner, and B. J. Streibel. 1996. *The team handbook,* 2nd ed. Madison, WI: Joiner Associates.

Summers, Donna C. S. 1997. *Quality.* Upper Saddle River, NJ: Prentice-Hall.

Uhlfedler, H. 1997. Ten critical traits of group dynamics. *Quality Progress* 30(4): 69–72.

Walton, M. 1986. *The Deming management method.* New York: Putnam.

Xerox Corporation. 1986. *Leadership through quality: Interactive skills workbook.* Stamford, CT: Xerox Corporation.

chapter

8

PROJECT MANAGEMENT DOCUMENTATION AND COMMUNICATIONS

"A horse, a horse! My kingdom for a horse," cried Shakespeare's Richard III. Although most of us no longer need horses to travel, we need good documentation for successful projects, lest we find ourselves crying, "Good documentation, good documentation! My career for good documentation."

This chapter is organized into two sections—Project Documentation and Project Communications. Project documentation is stressed because it is often neglected and because there are many other fine resources for project communication (i.e., reports and presentations), such as *A Beginner's Guide to Technical Communication,* by Anne Eisenberg (1998).

Project Documentation

Why are project documentation and record keeping important? To answer this question you only have to think about a time you had to pick up the pieces from a team member and pull them together into a report or presentation. Was everything easy to follow and understand, or did you have to fill in or reconstruct much of the work?

Problem solving, especially if it involves developing a computer program, is particularly susceptible to gaps in documentation. Think about how well you've documented projects you've been involved in. Did you insert lots of comments in your programs or spreadsheets to let the next user know what you'd done and why?

Many of the problem-solving and program-writing assignments students hand in lack sufficient documentation; thus, the faculty member cannot assess whether the procedures are correct (even if the answer isn't). The lack of documentation makes it very difficult for faculty to grade the report.

Project documentation is important because, as Leifer's (1997) research at the Stanford University Center for Design Research indicates, "all design is redesign" and the more information and insight you provide those who follow you, the better job they will be able to do. Of course, this works for you, too, when you follow up on someone else's work.

> **REFLECTION** Take a few minutes to think about the types of project records that should be kept. Make a list. Next, think about the characteristics of good records. List several attributes of good records. Compare your list with those of other team members.
>
> What did you come up with? How easy or hard was it for you to think about types of documents and the characteristics of good documentation?

Many engineers prefer to focus on solving the problem, developing the product, or just getting the job done rather than on documentation, which they see as a necessary (and often unpleasant) burden.

For comparison, here are lists that were developed by the participants in the Minnesota Department of Transportation's Project Management Academy:

Types of Records

Formal	Informal
Specifications	Survey notes
Drawings	Inspection reports
Schedules	Photographs/videotapes
Budgets	Notes—personal and meeting
Contracts	Incident reports
Change orders	Telephone/e-mail/fax memos and notes
On-site log—date/weather/time/personnel/equipment	Commitment logs

Characteristics of Good Records

Easily accessible	Consistent format
Thorough—date and time, client	Secure
Organized and legible	Cost effective
Comprehensive—table of contents	Flexible
Right media	

Nearly 200 forms were recently compiled by the Project Management Institute (PMI) in a book titled *Project Management Forms* (Pennypacker, Fisher, Hensley, and Parker, 1997). PMI members shared their forms, checklists, reports, charts, and other documents to help readers get started or to improve their current documentation. More information is available on the PMI Web site (www.pmi.org).

JOURNALS AND NOTES

Notebooks and journals are terrific ways to document work for your own personal use, and there are many examples of their significance in patent applications or even Nobel Prize considerations. Think, for example, of how the notebooks of Charles Kettering (inventor of auto electric cash register and automobile ignition systems) and Shockley, Brattain, and Bardeen (inventors of the transistor) helped establish intellectual property rights. Also consider Bill Gates recently paid $30 million for Leonardo da Vinci's *Codex Leicester,* one of da Vinci's surviving journals.

Students in ERG 291, a freshman design course at Michigan State University, are required to keep a laboratory notebook/academic journal. The box "Academic Journal" includes a description of the documentation requirements.

Chapter 7 of *Understanding Engineering Design* (Birmingham, Cleland, Driver and Maffin, 1997), Information in design, begins with the following sharply focused statement on the importance of information: "The raw material of the design process is information, and therefore the designer's principal skill is one of information handling" (p. 108). The authors stress five categories of action that operate on information:

1. Collection
2. Transformation
3. Evaluation
4. Communication
5. Storage

Furthermore they stress that these categories are used at all stages of the design process.

One of the challenges to personal notebooks and journals is that they are not easily shareable. Lack of easy access to others' work and thinking makes for considerable problems in team-based project work. Larry Leifer and his colleagues at Stanford have devised an electronic notebook that is accessible on the World Wide Web. The Personal Electronic Notebook with Sharing (PENS) (Hong, Toye, and Leifer, 1995) supports and implements Web-mediated selective sharing of one's working notes. Electronic mail is another common form of sharing thinking about projects. Numerous software products, such as Lotus Notes, provide a means for sharing information electronically. Electronic calendars and personal data assistants (PDAs) often have features for jointly scheduling meetings by viewing others' calendars. PDAs and electronic calendars also provide excellent means for keeping records. The image in Figure 8.1 is from the desktop interface of a Palm Pilot electronic calendar.

ACADEMIC JOURNAL

What Is a Journal? A journal is a place to practice writing and thinking. It differs from a diary in that it should not be merely a personal recording of the day's events. It differs from your class notebook in that it should not be merely an objective recording of academic data. Think of your journal rather as a personal record of your educational experience in this class. For example, you may want to use your journal while working on a design project to record reflections on the class.

What to Write. Use your journal to record personal reactions to class, topics, students, teachers. Make notes to yourself about ideas, theories, concepts, problems, etc. Record your thoughts, ideas, and readings; argue with the instructor; express confusion; and explore possible approaches to problems in the course. Be sure to include (1) critical incidents that helped you learn or gain insight, and (2) distractions that interfered with your learning.

When to Write. Try to write in your journal at least three or four times a week . . . aside from classroom entries. It is important to develop the *habit* of using your journal even when you are not in an academic environment. Good ideas, questions, etc., don't always wait for convenient times for you to record them. *[A man would do well to carry a pencil in his pocket and write down the thoughts of the moment. Those that come unsought for are the most valuable and should be secured because they seldom return. Francis Bacon (1561–1626)].*

How to Write. You should write in a style that you feel most comfortable with. The point is to think on paper without worrying about the mechanics of writing. The quantity you write is as important as the quality. The language that expresses your personal voice should be used. Namely, language that comes natural to you.

Suggestions:

1. Choose a notebook you are comfortable with. A 8½" x 11" hardback bound book with numbered pages would be a good choice.
2. Date each entry including time of day.
3. Don't hesitate to write long entries and develop your thoughts as fully as possible.
4. Include sketches and drawings.
5. Use a pen.
6. Use a new page for each new entry.
7. Include both "academic" and "personal" entries; mixed or separate as you desire.

Interaction—Professor. I'll ask to see your journal at least twice during the term: I'll read selected entries and, upon occasion, argue with you or comment on your comments. Mark any entry that you consider private and *don't want me to read* and I'll gladly honor your request. A *good* journal will contain numerous long entries and reflective comments. It should be used regularly.

Interaction—Correspondent. Choose a fellow student in your close collection of friends to read and respond to your journal entries.

Conclusions. Make a table of contents of significant entries. At the end of the semester write a two-page summary. In addition, submit an evaluation of whether the journal enhanced or detracted from your learning experience. Was it worth the effort?

SOURCE: Adapted from T. Fulwiler. 1987. *Teaching with writing.* Portsmouth, NH: Boynton/Cook. Revised by B. S. Thompson in consultation with K. A. Smith and R. C. Rosenberg, 1998.

Notice that in addition to having several calendar options, it also has an address book, a to-do list, a memo pad, and an expense section. PDAs such as the Palm Pilot provide for portability, backing up, and electronic sharing. In addition, all of the sections may be quickly searched, which makes for easy information. They will likely become an essential tool for project managers. Chapter 9 discusses computer-based tools further.

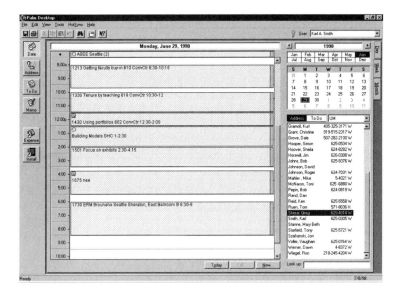

Figure 8.1 Palm Pilot PDA desktop view

PROJECT COMMUNICATIONS

Presenting your ideas to others both in written and oral form is essential to successful projects. You may, however, feel a combination of excitement and anxiety about report writing and public presentation.

Some of the best advice I ever got on presentations is to know your audience; know your objective; and be simple, concise, and direct. This brief section follows that advice.

Communications need to be tailored to the recipients whether they be your colleagues on the team, your supervisor, or your client. Common ways to tailor communications are to learn more about the recipients through surveys, interviews, or informal conversations.

REFLECTION Take a minute to reflect on some possible objectives in a presentation or report. Try to think beyond "Because it's an assignment in this class." As much as possible tie your list to actual experiences that you've had. Discuss these with other team members.

One of the best references on simple, concise, and direct writing is Strunk and White's (1979) *Elements of Style*. One of my favorites is Williams's (1997) *Style: Ten Lessons in Clarity and Grace*.

Getting Started, Keeping Going

A practice commonly called freewriting may help you get over the barrier that many of us face when starting to write a report or prepare for an oral presentation. A helpful implementation of freewriting is Natalie Goldberg's advice for "writing practice" in her book *Writing Down the Bones: Freeing the Writer Within* (1986):

1. *Keep your hand moving.* (Don't pause to reread the line you have just written. That's stalling and trying to get control of what you're saying.)
2. *Don't cross out.* (That is editing as you write. Even if you write something you didn't mean to write, leave it.)
3. *Don't worry about spelling, punctuation, grammar.* (Don't even care about staying within the margins and lines on the page.)
4. *Lose control.*
5. *Don't think. Don't get logical.*
6. *Go for the jugular.* (If something comes up in your writing that is scary or naked, dive right into it. It probably has lots of energy.)

Revising and Refining

After you complete a draft of your paper or presentation I encourage you to share it with others. Many papers and reports must undergo a formal review process, but I recommend that you offer your work for informal review with a colleague or friend. Although it's hard for many of us to share our work with others before we feel it is ready, we can often save a lot of time and get a much better product by asking others for comments.

At some stage in the process, of course, you must "freeze the design" and submit the report or give the presentation. Be sure to solicit feedback (in a constructive mode), reflect on it, and consider changes for the next time you write a report or give a presentation.

Leaving an excellent paper and electronic record is likely to increase in importance in projects and project management. Now is the time to develop skills and strategies for effective documentation. Similarly, communicating effectively both orally and in writing will continue to be extremely important for project success.

Questions

1. What is your experience documenting group projects? Is it a routine activity? If so, describe examples of "excellent documentation." If not, consider how can you build the development of good documentation into the ongoing process of project work.

2. What are some of the types of records that must be maintained for projects? What are the characteristics of good records?
3. What is your experience with electronic documentation? What are the advantages and disadvantages of electronic records (compared with paper records)?
4. Describe the characteristics of good presentations. Are good presentations the norm in your experience? Why or why not?
5. Describe your experiences keeping an academic journal. What are some of the heuristics that helped make it effective for you?
6. Learning to become a better writer and presenter requires effort and practice. How are you planning to improve your writing and presenting skills? What are some of your favorite resources?

REFERENCES

Eisenberg, A. 1998. *A beginner's guide to technical communication.* New York: WCB/McGraw-Hill.

Goldberg, N. 1986. *Writing down the bones: Freeing the writer within.* Boston: Shambhala.

Hong, J.; G. Toye; and L. Leifer. 1995. Personal Electronic Notebook with Sharing. In *Proceedings of the IEEE Fourth Workshop on Enabling Technologies: Infrastructure for Collaborative Enterprises (WET ICE).* Berkeley Springs, WV.

Leifer, L. 1997. A collaborative experience in global product-based-learning. National Technological University Faculty Forum. November 18, 1997.

Pennypacker, James S.; Lisa M. Fisher; Bobby Hensley; and Mark Parker. 1997. *Project management forms.* Newton Square, PA: Project Management Institute.

Strunk, W. Jr., and E. B. White. 1979. *The elements of style,* 3rd ed. New York: Macmillan.

Williams, J. M. 1997. *Style: Ten lessons in clarity and grace.* New York: Longman.

chapter

PROJECT MANAGEMENT SOFTWARE

"The more time we spend on planning a project, the less total time is required for it. Don't let today's busywork schedule crowd planning time out of your schedule."

Edward Bliss, *Getting Things Done.*

A wide variety of software tools is available to help the project manager and project team members accomplish their goals. These range from personal information managers (PIMs)—which include electronic calendars, address books, to-do lists, memo pads, and sometimes expense reports—to full-fledged project management programs that do scheduling, resource leveling, tracking, and so on.

PERSONAL INFORMATION MANAGERS

Personal information managers help project managers keep track of appointments, critical deadlines, notes, and expenses. Many provide for access to calendars over a network or over the Internet, a feature that makes it easy to schedule meetings. Most calendars can be synchronized between personal data assistants (PDAs), handheld devices such as the Palm Pilot, and portable computers that make it possible to easily take the information into the field. See Chapter 8 for a sample screen from the Palm Pilot desktop. Paper calendars and planners, although inexpensive, cannot be backed up easily (except by photocopying), nor can the information be shared with others very easily (which has its advantages).

INDIVIDUAL REFLECTION   What type of calendar or planner are you currently using? Is it a small paper datebook or a leather-bound three-ring binder? What are the principal uses that you make of your planner?

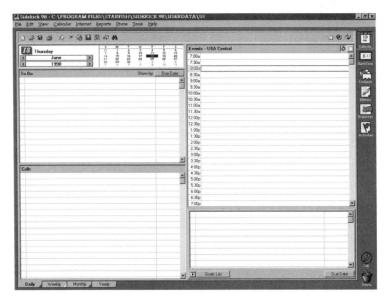

Figure 9.1   Sidekick PIM, desktop view

An example from the daily calendar view from the Sidekick personal information manager is illustrated in Figure 9.1.

Personal information managers and personal data assistants are mainly used to manage time, priorities, and contacts. They help project managers attend to the details that are crucial for successful teamwork and project management.

PROJECT MANAGEMENT SOFTWARE

Comprehensive project management software such as Microsoft Project is used on complex projects to accomplish goals and complete projects on time, within budget, and at a level of quality that meets the client's expectations. The basic functions of the critical path analysis aspect of these programs is summarized in Chapter 6, where the CritPath program is featured.

The two most common views used by commercial project management software are the Gantt chart and the precedence network chart, sometimes referred to as a program evaluation and review technique (PERT) chart. Figures 9.2 and 9.3 show examples of the Gantt chart and PERT chart views, respectively, from Microsoft Project.

Examples of the Gantt chart and PERT chart views from the Primavera software program are shown in Figures 9.4 and 9.5, respectively. These Primavera views show the activity detail for a highlighted activity; this allows the project manager to quickly get lots of detailed information on any activity, which makes it easy to track, manage, and update information.

PROJECT MANAGEMENT SOFTWARE

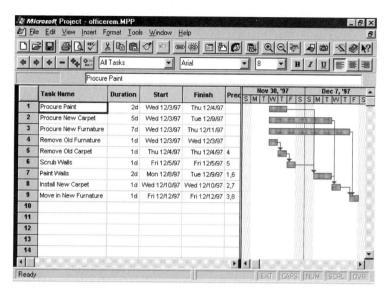

Figure 9.2 Microsoft Project—Gantt chart

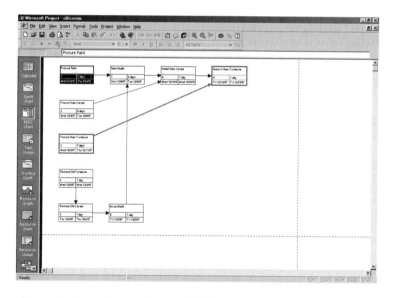

Figure 9.3 Microsoft Project—PERT chart

Microsoft Project and Primavera are the two most widely used project management software packages. In a survey regarding project management tools by Fox and Spence (1998), 48 percent of the respondents reported using Microsoft Project and 14 percent reported using Primavera.

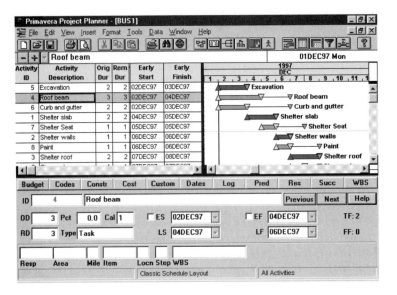

Figure 9.4 Primavera—Gantt chart

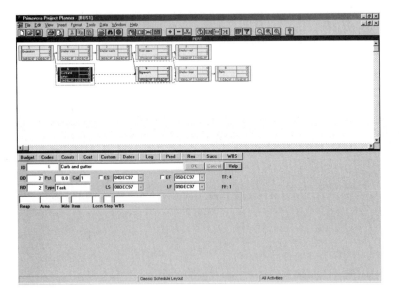

Figure 9.5 Primavera—PERT chart

Pollack-Johnson and Liberatore (1998) reported similar figures—nearly 50 percent for Microsoft Project and 21 percent for Primavera—and provided extensive information on how these packages are used. The median size of projects included in this study was a little over 150 activities, and the median number of resources was 16. A high percentage of respondents said they regularly update the information, and about 62 percent of the respondents said they use resource scheduling/leveling.

The Project Management Institute recently launched a major project management software survey, available both in print and on CD-ROM (Cabanis, 1999). The survey

- Compares and contrasts the capabilities of a wide variety of project management tools.
- Provides a forum for users and vendors to meet and match requirements and possibilities.
- Prompts vendors to become more responsive to customer needs.
- Prompts users to create a method for software tool selection within their own companies.
- Categorizes software tools into six areas of functionality aligned with the knowledge areas of PMI's *Guide to the Project Management Body of Knowledge (PMBOK™ Guide):* scheduling, cost management, risk management, human resources management, communications management, and process management.

As with the use of all software tools, it is important that project management software serve and not enslave the project manager. Also, if you invest time and money in commercial project management software, you should use it to organize and manage your projects and not simply to write reports. Lientz and Rea (1995) offer the following suggestions for using project management software:

1. Set up the basic schedule information: name of project file, name of project, project manager; input milestones, tasks and their estimated duration, interdependencies between tasks; input resources for each task.
2. Periodically update the schedule by indicating tasks completed, delayed, and so forth, as well as changes in resources.
3. On an as-needed basis, perform what-if analysis using the software and data.

Uses of project management software include *reporting* (use schedule to produce graphs and tables for meetings); *tracking* (log project work and effort in terms of completed tasks); *analysis* (perform analysis by moving tasks around, changing task interdependencies, changing resources and assignments, and then seeing the impact on the schedule); *costing and accounting* (assign costs to resources); and *timekeeping* (enter the time and tasks worked on by each member of the project team.).

Unofficial reports indicate that over 1 million copies of Microsoft Project have been sold. That's a lot of people scheduling projects. Advertisements for civil engineering positions often include familiarity with project management software, especially Primavera. Lots of books, short courses, and multimedia training programs are available to help you learn to use these tools. Some of the books I've found useful are included in the references (Day, 1995; Lowery, 1994; Marchman, 1998). This is a rapidly changing area, so to keep current you should stay tuned to resources such as the Project Management Institute, especially via its Web site (www.pmi.org). If you are not familiar with browsing and searching the Internet, a good resource is *Introduction to the Internet for Engineers* (Greenlaw and Hepp, 1999).

PROJECT MANAGEMENT AND THE WORLD WIDE WEB

Project management, like many other things—bookstores, newspapers, computer suppliers—has developed a Web presence. As mentioned above, the Project Management Institute has a very thorough Web site and provides access to the *Guide to the Project Management Body of Knowledge (PMBOK™ Guide)*. A few project management Web sites I have found useful are the following:

Project Management Institute—http://www.pmi.org

HMS Project Management Articles—
http://www.hmssoftware.ca/articles/art_list.html

ALLPM: The Project Manager's Home Page—http://allpm.com/index.htm

Project Control Tower—http://www.4pm.com/

WELCOM: Project Management for a Changing World—
http://www.welcom.com/

NewGrange Center for Project Management—http://www.newgrange.org

Project Management Leadership Group—http://www.pmlg.org

Web-enabled project management is gaining momentum and will probably change some businesses, as it is currently doing in the construction industry (Roe and Pfair, 1999; Doherty, 1999). Web-enabled project management couples both the communication aspects (e-mail, fax, voice and multimedia, intranet, extranet) with the project management aspects (scheduling, document and file management, project administration, job photos, job cost reports, and project status reports). Doherty (1999) cites several reasons for using a project extranet:

- Fewer communication errors between project team members.
- Up-to-the-minute intelligence on all the decisions and collective information related to a project.
- Less expense for messengers, couriers, copying, and blueprints.
- Customized sites for each project and customized access for each user.
- Security.

Since project management is about planning, scheduling, monitoring, and controlling, projects that have a central project file located at a Web site rather than in a project notebook (or in the project manager's head) have enormous benefits. The challenge involves moving from our comfort zones of familiar practice and learning new tools and approaches.

Questions

1. Describe the advantages and disadvantages of different calendar/planner formats—pocket planner, three-ring binder, pocket electronic organizer, and computer-based personal data assistant.
2. What are the major types of project management software? What are their common uses?
3. What are the advantages and disadvantages of the Gantt chart and the precedence network (or PERT) charts available in commercial project management software?
4. How could you apply Lientz and Rea's suggestions for using project management software to a project you're currently involved in?
5. Check out project management on the World Wide Web. Keep a record of your findings in a journal.

References

Cabanis, Jeannette. 1999. *Project management software survey.* Newton Square, PA: Project Management Institute.

Day, Peggy. 1995. *Microsoft Project 4.0: Setting project management standards.* New York: Van Nostrand Reinhold.

Doherty, Paul. 1999. Site seeing. *Civil Engineering* 69(1): 38–41.

Feigenbaum, Leslie. 1998. *Construction scheduling with Primavera project planner.* Upper Saddle River, NJ: Prentice-Hall.

Fox, Terry L., and J. Wayne Spence. 1998. Tools of the trade: A survey of project management tools. *Project Management* 29(3): 20–27.

Greenlaw, Raymond, and Ellen Hepp. 1999. Introduction to the Internet for engineers. New York: McGraw-Hill.

Lientz, Bennet, and Kathryn Rea. 1995. *Project management for the 21st century.* San Diego: Academic Press.

Lowery, Gwen. 1994. *Managing projects with Microsoft Project 4.0.* New York: Van Nostrand Reinhold.

Marchman, David A. 1998. *Construction scheduling with Primavera project planner.* Albany, NY: Delmar.

Pollock-Jackson, Bruce, and Matthew J. Liberatore. 1998. Project management software usage patterns and suggested research directions for future development. *Project management* 29(2): 19–28.

Roe, Arthur G., and Matthew Phair. 1999. Connection crescendo. *ENR* (May 17): 22–26.

chapter

10

WHERE TO GO FROM HERE

Projects and teams are going to be with you for the rest of your life, no matter what profession you eventually work in. They are already prevalent in engineering, medicine, law, and most areas of business and industry. Even if you become a college professor, you will probably be involved in projects and teams, especially on research projects and with your graduate students. Now you have made a start at learning how to effectively participate in project management and teamwork. In addition to learning how to participate effectively, I hope you've developed skills for managing and leading a team. There are lots of additional resources available, and I hope you will continue to read about project management and teamwork. More important, I hope you will talk with colleagues (fellow students and faculty) about project management and teamwork. If you aspire to become a project manager, I encourage you to check out the Project Management Institute. PMI has a special student membership rate, and your membership in this organization will help connect you with project management professionals. Most professional organizations, such as American Society of Civil Engineers (ASCE), American Society of Mechanical Engineers (ASME), and Institute for Electrical and Electronic Engineers (IEEE), have a division that emphasizes engineering management. Check these out as you become a student member of the professional organization in your discipline.

Periodically reflecting on your experiences, writing down your reflections (as I have asked you to do throughout this book), processing them alone and with others, and reading and studying further will help ensure that your project and team experiences become ever more constructive. A sustained effort will ensure that you continue to learn and grow.

If you are in a team or project situation that is not working well, rather than just endure it and hope it will pass quickly, try some of the ideas in this book for improving the team or project. Suggest that the team members discuss how effectively they are working. For example, suggest a quick, individually written plus/delta processing exercise to survey the team. Successful project work and teamwork do not just magically happen; each takes continual attention not only to the task but also to how well the group is working. And this is work. The satisfaction and sense of accomplishment that come from effective teamwork and project work are worth the effort. There are just so many things that can't be accomplished any other way. The more you learn during your undergraduate years, the easier it will be for you after graduation. Paying attention to these skills now will save you what previous generations of engineering graduates have had to endure—learning project management and teamwork skills on the job in addition to all the other complex things they had to learn.

As you work with this book and the ideas and strategies for effective project management and teamwork, please think about what else you need to know. Develop a learning and teaching plan for yourself and your project and team members. The resources listed at the end of each chapter in this book only barely scratch the surface of all the resources that are available. Check out a few of them. Add your own favorites. Share your list with colleagues. As mentioned earlier several students I've worked with have found Stephen Covey's *Seven Habits of Highly Effective People* very helpful (see Chapter 5). Covey's book is also a perennial best-seller. The earlier you learn the skills and strategies for effective project work and teamwork, the more productive you will be and the easier life will be for you later. Start now.

Although the up-front goal of this book is to facilitate the development of project management and teamwork skills in engineering students, the deeper goal is to change the climate in engineering courses and programs from competitive or negative interdependence to cooperative or positive interdependence; from suspicion, mistrust, and minimal tolerance of others to acceptance, trust, and valuing others; from an egocentric "What's in it for me?" to a communal "How are we doing?"; from a sense of individual isolation and alienation to a sense of belonging and acceptance. I recognize that these are lofty goals, but until we take more responsibility not only for our own learning and development but also for the learning and development of others, we will not benefit from synergistic interaction.

If you find that project management and teamwork, and perhaps even leadership, are of great interest to you, then you may want to read some of the business magazines such as *Business Week* or the *Harvard Business Review,* and perhaps even my favorite, *Fast Company.* Check out your local bookseller or Internet bookstore for some spare-time reading on these topics. You'll find an enormous literature available.

If you're more interested in project management and teamwork specifically within engineering and technology, then I suggest that you look into some of the books and video documentaries on projects, such as Karl Sabbagh's work—*Skyscraper* and *21st Century Jet.*

The first project management book I encountered, many years ago, was written by Harold Kerzner. Kerzner was one of the first advocates of a systems approach to project management. I'd like to give the last word to him, a list of 16 points to project management maturity (and you know by now that I really like lists), from the latest edition (1998) of his tome on project management:

1. Adopt a project management methodology and use it consistently.
2. Implement a philosophy that drives the company toward project management maturity and communicate it to everyone.
3. Commit to developing effective plans at the beginning of each project.
4. Minimize the scope changes by committing to realistic objectives.
5. Recognize that cost and schedule management are inseparable.
6. Select the right person as the project manager.
7. Provide executives with project sponsor information, not project management information.
8. Strengthen involvement and support of line management.
9. Focus on deliverables rather than resources.
10. Cultivate effective communication, cooperation, and trust to achieve rapid project management maturity.
11. Share recognition for project success with the entire project team and line management.
12. Eliminate nonproductive meetings.
13. Focus on identifying and solving problems early, quickly, and cost-effectively.
14. Measure progress periodically.
15. Use project management software as a tool—not as a substitute for effective planning or interpersonal skills.
16. Institute an all-employee training program with periodic updates based upon documented lessons learned.

REFERENCES

Kerzner, Harold. 1998. *Project management: A systems approach to planning, scheduling, and controlling,* 6th ed. New York: Van Nostrand Reinhold.

Sabbagh, Karl. 1996. *Twenty-first century jet: The making and marketing of the Boeing 777.* New York: Scribners.

Sabbagh, Karl. 1991. *Skyscraper: The making of a building.* New York: Penguin.

INDEX

NAME & SUBJECT

A

Accountability, 18
Accreditation Board for Engineering and Technology, 3, 6
Ackoff, Russell L., 6, 10
Activity ranking, 8
Adams, James L., 3, 10
Age of Project Management, 57
American Society of Civil Engineers, 111
American Society of Mechanical Engineers, 111
Analysis, 107
Arizona State University's Introduction to Engineering Design, 26
Austin, Nancy, 93
Average of members' opinions, 30

B

Backward pass, 68–69
Bacon, Francis, 98
Bahill, A. Terry, 3, 10
Bardeen, John, 97
Beginner's Guide to Technical Communication (Eisenberg), 95
Behavioral changes, 56
Belgard, William, 36, 41
Bellamy, Lynn, 24, 25, 26, 41
Bennis, Warren, 18, 19, 20, 88, 92
Biederman, Patricia W., 18, 19, 20, 88, 92
Blake, R. R., 33, 41
Bleloch, A. L., 4, 64, 71, 80
Bliss, Edward, 103
Boeing Company
    code of cooperation, 25
    employer checklist, 5–6
    group effectiveness, 83
    777 project, 45–47
Bowles, J., 91, 92
Brassand, Michael, 19, 20, 91, 92
Brattain, Walter H., 97
Brooks, Frederick P., Jr., 62, 63
Brown, John Sealy, 92

Brown, S. M., 21
Browne, M. Neil, 16, 20
Bucciarelli, Louis, 3, 10
Burlington Northern International Transport Team, 17
Burton, Lawrence, 8, 10
Business Week, 112
Byrne, J. A., 53–54, 63

C

Cabanis, Jeannette, 18, 20, 107, 109
Case studies
    conflict management, 40
    project scheduling, 80
Chapman, William L., 3, 10
Cherbeneau, Jeanne, 18, 20
Chicago Bulls, 18
Churchman, C. West, 6, 10
Classroom learning group, 16
Cleary, Barbara A., 91, 93
Cleese, John, 83, 92
Cleland, D. I., 44, 51
Client acceptance, 45–46, 60
Client consultation, 60
Code of cooperation, 25–26
Codex Leicester (Leonardo), 97
College teaching, 55
Communication, 60; *see also* Project communication
Communication skills, 26–27
    of project manager, 58
Compromise, 33
Computer-based project management software, 78
Concurrent engineering, 5
Condit, Phil, 46
Conflict
    definition, 32
    steps in negotiating, 34
Conflict management, 32–35
    case, 40
    tactics, 87–88

Conflict strategies, 33–34
Confrontation, 33
    guidelines for, 34
Consensus, 31
Constructive feedback, 84–86
Continual evaluation, 90
Controlling, 58
    role of project manager, 61
Cooperative learning group, 16
Coping skills, 58
Cost-effectiveness of models, 4
Cost factors, 45
Costing and accounting, 107
Covey, Stephen R., 61, 63, 112
Crawford-Mason, Clare, 91, 92
Creative listening, 27
Critical listening, 27
Critical path, 69
Critical path method, 5, 43
    backward pass, 68–69
    example, 72–74
    forward pass, 68
    Gantt charts, 70–74
    for project scheduling, 67–74
    summary of, 72
Critical success factors, 60
CritPath software program, 70–74, 104
Culp, G., 50, 51

D

Day, Peggy, 107, 109
Death March Projects (Yourdon), 43
Decision by authority, 30
Decision making
    methods, 30–31
    in teams, 29–32
Decisions, effective, 31
Delays, 90
Delehanty, Hugh, 18, 21
Deming, W. Edwards, 5, 91, 92
Denker, G. R., 43, 52
Dertouzos, M. L., 54, 55, 63

Design; *see* Engineering design
Directing role of project manager, 61
Distributed actions approach, 23–24
Diversity, 17–18
Dobyns, Lloyd, 91, 92
Doherty, Paul, 108, 109
Donnellon, Anne, 86–88, 92
Dressler, C., 81, 93
Drucker, Peter F., 56, 63
Duguid, Paul, 92
Duncan, William R., 49, 51

E

Early-start-early finish, 68
Effective teams, 13–14
    characteristics, 18–19
Eisenberg, Anne, 95, 101
Eisner, H., 50, 51
Elements of Style (Strunk & White), 99
Employer checklists, 5–6
Engineering, 1–2
    behavioral changes, 56
    management in, 5
    modeling in, 3–5
    professional organizations, 111
    project management, 8–9
    systems approach, 6–8
    teamwork in, 5–6, 8–9
Engineering design, 2
    as social process, 3
Engineering education, suggested changes in, 53–55
Engineering heuristics, 4–5
Englund, Randall L., 50, 52
Excellence in Engineering (Roadstrum), 8–9
Expert member, 30

F

Fast Company, 112
Feedback, 60
Feigenbaum, Leslie, 109
Ferguson, Eugene S., 3, 10
Fisher, Kimball, 36, 41
Fisher, Lisa M., 96, 101
Floats, 69
Forcing, 33
Ford, Henry, 13
Ford Motor Company
    code of cooperation, 25–26
    motto, 91

Fortune, 50
Forward pass, 68
Foster, Sallie, 24, 25, 26, 41
Fox, Terry L., 105, 109
Frame, J. D., 50, 51
Free floater, 69
Freewriting, 100
Frontiers in Education Conference, 54
Fulwiler, T., 98
Futernick, Jennifer, 16

G

Gantt charts, 5, 70–74, 104–106
Gates, Bill, 97
Gaynor, G. H., 55, 63
Gershenfeld, Matti K., 24, 41
Goldberg, David E., 36, 41
Goldberg, Natalie, 100, 101
Grady, Robert B., 50, 52
Graham, Robert J., 50, 52
Greenlaw, Ray, 108, 109
Group accountability, 18
Group effectiveness, 83–86
    constructive feedback, 84–86
Group maintenance roles, 24
Group norms, 24
Group performance, 15
Group processing, 19
Group projects, 5
Groups
    compared to teams, 17, 86
    nominal or real, 88
Group task roles, 24
Guide to the Project Management Body of Knowledge, 107, 108

H

Hackman, J. R., 19, 20, 36, 41
Hammond, J., 91, 92
Hancock, J. C., 2, 10
Hapgood, Fred, 3, 10
Hargrove, Robert, 18, 20
Harvard Business Review, 56, 112
Haynes, M. E., 89, 93
Hensley, Bobby, 96, 101
Hepp, Ellen, 108, 109
Heuristics, 4–5, 49
High-performance cooperative learning group, 16–17
High-performing teams, 14
Hoepner, Ken, 16–17

Hong, J., 97, 101
How to Model It (Starfield et al.), 71

I

Identification, 86
Individual accountability, 18
Institute for Electrical and Electronic Engineers, 111
Interactive Skills Coding Worksheet, 86–87
Interdependence, 86
Introduction to Engineering Design, 26

J

Jackson, Phil, 18, 21
Jay, A., 92
Johnson, David W., 19, 21, 30–31, 39, 41
Johnson, Frank P., 19, 21, 41
Johnson, Roger T., 18, 21, 30–31, 39, 41
Joiner, Brian L., 19, 21, 32, 36, 41, 82, 93
Jordan, Brigitte, 7, 10
Jordan, Michael, 18
Journals, 97–99

K

Katzenbach, Jon R., 14, 16, 17, 18, 19, 21, 36, 41, 55–56, 63, 88, 93
Keeley, Stuart, 16, 20
Kerzner, Harold, 44, 47, 51, 52, 112, 113
Kettering, Charles, 97
Kharbanda, O. P., 45, 46, 50, 52, 57, 58, 64
Kiser, Kenneth J., 91, 93
Knowledge workers, 56
Koen, Billy V., 2, 4, 9, 10
Kouzes, J. M., 27–28, 41

L

Langford, David P., 91, 93
Late start-late finish, 68–69
Laufer, A., 43, 52
Leader's Handbook (Scholtes), 29
Leadership
    behavioral commitment, 28–29
    characteristics, 27–28
    competencies of, 29
    distributed actions approach, 23–24
Leadership skills, 58
Learning principles, 7

Learning teams, 15–17
LeBold, William K., 8, 10
Leifer, Larry, 3, 10, 21, 96–97, 101
Leonardo da Vinci, 97
Lester, R. K., 54, 55, 63
Lewis, James P., 50, 52, 93
Liberatore, Matthew J., 106, 109
Lientz, Bennet, 50, 52, 57, 58, 64, 107, 109
Lippincott, S., 82, 93
Listening techniques, 27
Loparo, K. A., 11
Lotus Notes, 97
Lowery, Gwen, 107, 109

M

"Made in America" study, 53–54
Maintenance roles, 24
Majority control, 31
Management
    behavior change, 24
    in engineering, 5
Managing diversity, 17–18
Mantel, S. J., 50, 52
Marchman, David A., 107, 109
Massachusetts Institute of Technology, 54
Mastering the Art of Creative Collaboration (Hargrove), 19
McConnell, Steve, 62, 64
McKinsey & Company, 16
McNeill, Barry, 24, 25, 26, 41
Meetings
    conducting, 83
    guidelines for, 82
    major complaints about, 81–82
    process, 82–83
Meetings: Do's, Don't's, and Donuts (Lippincott), 82
Mental models, 7
Meredith, J. R., 50, 52
Michigan State University, 97
Microsoft Project, 104–107
Minnesota Department of Transportation Project Management Academy, 96
Minority control, 30
Modeling, 3–5
Models
    attributes of, 4
    and heuristics, 5
    of project management, 44
    project management heuristics, 49
Monitoring, 60

Mouton, J. S., 33, 41
Mythical Man-Month (Brooks), 62

N

Napier, Rodney W., 24, 41
Negotiation process, 88
Network diagrams, 5
New-paradigm companies, 53–54
Nicholas, J. M., 44, 50, 52
Nielsen, N. R., 11
Nominal teams, 88
Noncritical activities, 69
Notebooks, 97–99

O

Organizational skills, 58
Organizing Genius (Bennet & Biederman), 19
Organizing role of project manager, 60

P

Palm Pilot, 97, 98, 103
Papalambros, Panos Y., 3, 10
Parker, Linda, 8, 10
Parker, Mark, 96, 101
Pennypacker, James S., 96, 101
Performance, 45
Personal data assistant, 97, 103
Personal Electronic Notebook with Sharing, 97
Personal information managers, 103–104
Personal mastery, 7
Personnel selection, 60
PERT (program and evaluation review technique), 43
    charts, 104–106
Peters, Thomas J., 56–57, 64, 81, 93
Pfair, Matthew, 108, 109
Phillips, D. T., 49, 52
Pinto, Jeffrey K., 45, 46, 50, 52, 57, 58, 59, 60, 64
Pippin, Scottie, 18
Planning role of project manager, 59–60
Planning skills, 58
Pollack-Johnson, Bruce, 106, 109
Positive interdependence, 18
Posner, Barry Z., 27–28, 41, 58, 64
Potential teams, 14
Power differentiation, 87
Primavera software, 104–107
Problem solving, 95

Project communication, 99–100
Project documentation
    characteristics, 95–97
    journals and notes, 97–99
    types of records, 96
Project engineer, 8–9
Project evaluation
    continual, 90
    quality issues, 90–91
    questions for, 89
Project life cycle, 47–49
    project manager role
        controlling, 61–62
        directing, 61
        organizing, 60
        planning, 59–60
        staffing, 60
Project management, 5, 8–9
    changes in, 53, 56–57
    Kerzner's guidelines, 112–113
    models of, 44
    and World Wide Web, 108
Project Management Forms (Pennypacker et al.), 96
Project management heuristics, 49
Project Management Institute
    on-time finish survey, 43
    on project documentation, 96
    software survey, 107
    Web site study, 108
Project management software, 78
    main types, 104–107
    survey, 107
    uses, 107–108
Project managers
    behavioral changes, 56
    changing roles, 53–54
    role in project life cycle, 59–62
    skills for effectiveness, 58–59
Project mission, 60
Project monitoring
    group effectiveness, 83–86
    at meetings, 81–83
    team talk analysis, 86–89
Project planning, 49–50
Projects, 44–45
    bringing quality to, 90–91
    client acceptance, 45–46, 60
    constraints on success, 45–46
    cost considerations, 76–77
    critical success factors, 60
    delays and resource excess, 90
    keys to success of, 45–47

Projects—*Cont.*
   on-time completion, 43
   resource considerations, 74–75
Project scheduling, 65
   case study, 80
   critical path method, 67–74
   resource and cost considerations, 74–77
   work breakdown structure, 66–67
Project teams, 5
Promotive interaction, 18
Pseudo learning group, 15
Pseudo teams, 14
Purpose, 4

Q-R

Quality, contrasting views of, 90–91
Ranking work activities, 8
Ravindran, A., 49, 52
Ray, Michael, 7, 10
Rayner, Steven, 36, 41
Rea, Kathryn, 50, 52, 57, 58, 64, 107, 109
Real teams, 14, 88
Records, 96
Reference, 4
Relationships, 23–24
Reporting, 107
Resource considerations, 74–75
Resource excesses, 90
Richard III, 95
Rinzler, Alan, 7, 10
Roadstrum, W. H., 8–9, 11
Roberts, Harry V., 11
Rockefeller, John D., 23
Rodman, Dennis, 18
Roe, Arthur G., 108, 109
Rosenberg, R. C., 98
Rosenblatt, A., 11
Rothenberg, Jeff, 3, 11

S

Sabbagh, Karl, 47, 52, 112, 113
Sacred Hoops (Jackson & Delahanty), 18
Salami statistics, 58
Sashkin, Marshall, 91, 93
Schedule plans, 60
Scholtes, Peter R., 19, 21, 29, 32, 36, 41, 82, 84, 93
School of Engineering, 5
Schrage, Michael, 13–14, 18, 19, 21
Senad, G. L., 44, 45, 50

Senge, Peter, 7, 11
Seven Habits of Highly Effective People (Covey), 61, 112
Shakespeare, William, 95
Shared Minds (Schrage), 19
Shared vision, 7
Shenhar, A. J., 43, 52
Shina, S. G., 5, 11
Shockley, William B., 97
Siedner, C. J., 21
Skyscraper (Sabbagh), 112
Slack, 69
Smith, A., 50, 51
Smith, Douglas K., 14, 16, 17, 18, 19, 21, 36, 41, 55–56, 63, 88, 93
Smith, Karl A., 4, 18, 21, 54, 55, 64, 71, 80, 98
Smoothing, 33
Snead, G. L., 45, 52
Social distance, 87
Social process, 3
Software Project Survival Guide (McConnell), 62
Software tools
   personal information managers, 103–104
   for project management, 104–108
Solberg, J. J., 49, 52
Solow, R. M., 54, 55, 63
Spence, J. Wayne, 105, 109
Sperlich, Harold K., 13
Staffing role of project manager, 60
Stanford University Center for Design Research, 3, 96
Starfield, A. M., 4, 64, 71, 80
Statement of work, 49
Streibel, Barbara J., 19, 21, 32, 36, 41, 82, 93
Striving for Excellence in College (Browne & Keeley), 16
Strunk, W., Jr., 99, 101
Style: Ten Lessons in Clarity and Grace (Williams), 99
Summers, Donna C. S., 91, 93
Superconducting Supercollider, 45
Sympathetic listening, 27
Systems, 6
Systems approach, 6–8
Systems thinking, 6–7

T

Task roles, 24
Tasks, 23–24

Taylor, James, 27, 41, 50, 52
Team-building skills, 58
Team charter, 26
Team development, 32
Team effectiveness, 18–19
Team Handbook (Scholtes et al.), 29, 36, 82
Team learning, 7
Team performance, advice on, 19
Teams; *see also* Learning teams
   categories of, 14
   compared to groups, 17, 86
   effective, 13–14
   literature on, 19
Team talk analysis, 86–89
Teamwork, 5–6, 8–9
   challenges and problems, 35–36
   importance of diversity, 17–18
   literature on, 19
Teamwork skills, 18–19
   communication skills, 26–27
   conflict management, 32–35
   decision making, 29–32
   group norms, 24
   leadership, 27–29
   tasks and relationships, 23–24
Technical tasks, 60
Technological skills, 58
Thompson, B. S., 98
Time factors, 45
Time-keeping, 107
Time-scaled network, 70
Today's Engineer, 54–55
Top management support, 60
Total float, 69
Total quality management, 5
Toye, G., 97, 101
Tracking, 107
Traditional classroom learning group, 16
Tribus, Myron, 5, 11
Troubleshooting, 60
21st Century Jet (Sabbagh), 46–47, 112

U-V

Uhlfedler, H., 84, 93
Understanding Engineering Design (Birmingham et al.), 97
Virtual representations, 5

W

Waller, Alisha A., 54, 55, 64
Walton, M., 91, 93
Watson, G. F., 11

Weisbord, Marvin R., 7, 11
Welch, John F., 23
White, E. B., 99, 101
Widman, L. E., 11
Wilde, Douglass J., 3, 10, 19
Williams, J. M., 99, 101
Wisdom of Teams (Katzenbach & Smith), 19
Withdrawal, 33

Work activities, 8
Work breakdown structure, 49–50, 66–67
Workplace changes, 53–56
World Wide Web, 97, 108
Writing Down the Bones (Goldberg), 100
Writing skills, 99–100
Wycoff, J., 44, 45, 50, 52
Wymore, A. Wayne, 3, 10

X-Y

Xerox Corporation
    Interactive Skills Coding Worksheet, 86–87
    learning principles, 7
Yourdon, Ed, 43, 52